D1085244

BROADBAND WIRELESS ACCESS

THE KLUWER INTERNATIONAL SERIES
IN ENGINEERING AND COMPUTER SCIENCE

BROADBAND WIRELESS ACCESS

Benny Bing

Kluwer Academic Publishers
Boston/Dordrecht/London

Distributors for North, Central and South America:
Kluwer Academic Publishers
101 Philip Drive
Assinippi Park
Norwell, Massachusetts 02061 USA
Telephone (781) 871-6600
Fax (781) 871-6528
E-Mail <kluwer@wkap.com>

Distributors for all other countries:
Kluwer Academic Publishers Group
Distribution Centre
Post Office Box 322
3300 AH Dordrecht, THE NETHERLANDS
Telephone 31 78 6392 392
Fax 31 78 6546 474
E-Mail services@wkap.nl>

 Electronic Services <http://www.wkap.nl>

Library of Congress Cataloging-in-Publication

Bing, Benny.
  Broadband wireless access / Benny Bing.
    p. cm. -- (Kluwer international series in engineering and computer science ; SECS 578)
  Includes bibliographical references and index.
  ISBN 0-7923-7955-1
  1. Broadband communication systems. 2. Multimedia systems--Congresses. 3.
Computer network protocols. I. Title. II. Series.

  TK5103.4 .B48 2000
  621.382'12--dc21

                                                    00-059284

Printed on acid-free paper.

Printed in the United States of America

As always, to my mother

TABLE OF CONTENTS

Chapter 1   Overview of Wireless Networks

Chapter 2   Wireless Access Protocol Design

Chapter 3   Multiple Access Communications

Chapter 4   Fixed Allocation Access Protocols

Chapter 5   Contention Protocols

Chapter 6   Spread Spectrum Multiple Access

Chapter 7   Reservation Protocols

Chapter 8   Broadband Wireless Access Protocols

Chapter 9   A Generalized Broadband Wireless Access Protocol

Appendix   Queuing Theory Primer

"The wireless music box has no imaginable commercial value. Who would pay for a message sent to nobody in particular?"
— *David Sarnoff's associates in response to his urging for investment in radio, 1920s.*

Preface

The last century has seen the introduction of many types of wireless communication services which have become the cornerstone of modern telecommunications. Broadcast radio and television have dramatically changed our perception and understanding of the world we live in. The use of cellular phones has brought about the freedom to roam to different vicinities and yet communicate. More importantly, the success of cellular wireless service in support of vehicular and pedestrian users has created a pressing demand for such wireless service to be coupled with emerging broadband wireline infrastructure involving the Internet, thereby extending the benefits of multimedia services from the home and business environments, and heralding in a new era of anyone, anywhere, anytime, any media communications. As we start the new millennium, such exciting wireless communication technologies are already evolving and expanding at a phenomenal pace. Third-generation personal communication systems have been planned while high-speed wireless asynchronous transfer mode (ATM) networks and wireless Internet connectivity are the major focus of recent research efforts. These broadband networks aim to provide integrated, packet-oriented, transmission of text, graphics, voice, image, video, and computer data between individuals as well as in the broadcast mode.

In determining the performance of broadband wireless systems, many factors come into play. These factors are heavily dependent on the characteristics of the wireless channel such as signal fading, multipath distortion, limited bandwidth, high error rates, rapidly changing propagation conditions, mutual interference of signals, and the vulnerability to eavesdrop and unauthorized access. Moreover, the performance observed by each individual user in the network is different and is a function of its location as well as the location of other interacting users. Issues related to connection control and traffic management must also be addressed. The biggest challenge is to support real-time applications by providing bandwidth on demand (i.e., adequate quality of service) seamlessly across wireless and wireline networks.

One of the crucial elements of these advanced wireless systems is the access protocol, which defines the way a common bandwidth resource (the communication channel) is shared among contending users and hence, determines the overall performance of the system. For broadband wireless networks, the access protocol must ensure efficient and timely (on-demand) access to multirate applications with different communications requirements. In addition, the protocol is required to operate with the difficult constraints posed by moving users, dynamic variation in traffic patterns, and highly sensitive wireless links. Issues related to the connection control and traffic management of widely disparate traffic streams must also be addressed. The biggest challenge is to support real-time applications by providing bandwidth on demand (i.e., adequate quality of service) seamlessly across wireless and wireline networks.

Many access schemes currently deployed in mobile cellular networks employ a fixed frame structure. For instance, time division multiple access (TDMA) is a popular choice for several digital cellular systems such as GSM, IS-54, PDC and for wireless personal communication systems such as DECT, PHS, WACS. However, to cater for packetized multimedia traffic, the selection of a suitable TDMA frame structure is a not an easy task since it is unlikely that the exact mix of applications will either be known beforehand or remain stable. If time slots (that constitute a TDMA frame) are chosen to match the largest message lengths, slot times that are under-utilized by short messages must be padded out to fill up the slots. On the other hand, if shorter slot sizes are used, more overhead per message results. For some TDMA schemes, long frames are necessary in order to maximize the bandwidth utilization at high traffic load but this is done at the expense of increasing the delay at low load. To accommodate the greater levels of uncertainty created by the mixing of a wide assortment of applications, multiple access schemes supporting multimedia services should strive to achieve more flexible bandwidth sharing by allowing users to seize variable amounts of bandwidth on demand.

The access technique in spread spectrum and wideband code division multiple access (CDMA) networks usually refers to the ability of certain kinds of signals to coexist in the same frequency and time space with an acceptable level of mutual interference. The use of pseudorandom waveforms in a wireless network is motivated largely by the desire to achieve good performance in fading multipath channels and the ability to operate multiple links with pseudo-orthogonal waveforms using spread spectrum multiple access. A number of hybrid CDMA, multicarrier and orthogonal frequency division multiplexing (OFDM) schemes have been proposed for high-speed wireless communications. Multicarrier schemes that employ parallel signaling methods offer several significant advantages over conventional single carrier systems including protection against dispersive multipath channels and frequency-selective fading. With appropriate signal processing, these systems can achieve the equivalent capacity and delay performance of single carrier systems without the need for a continuous spectrum. Although the technology is promising, many technical challenges need to be overcome before spread spectrum-based systems can efficiently accommodate the high and variable bit rates demanded by broadband multimedia services.

The purpose of this book is to discuss the design and development of wireless access protocols with emphasis on how such protocols can efficiently support disparate classes of multimedia traffic. Besides a comprehensive introduction and survey to the evolution of wireless access protocols, many important protocols that are deployed or experimented in various broadband wireless environments (e.g., wireless ATM, broadband satellite networks, mobile cellular and personal communication systems, wireless local loops, wireless local area and home networks) are also covered.

I have attempted to make the book accessible to a wide audience. It serves as an excellent bridge between novice readers who wish to know more about the subject and advanced readers who are undertaking research in this area. While engineering aspects are discussed, the emphasis is on the physical understanding of access protocols, from basic operations to the latest innovations. To this end, mathematical treatment is kept to a minimum and most of the fundamental concepts are explained based on intuition and insights, supplemented by numerous illustrative figures. In addition, these concepts are expressed succinctly and concisely. Wherever possible, important references that make major contributions to the field are included for interested readers to investigate further. Latest updates and useful Web resources are also posted at the book's Web site. Readers are encouraged to send an email to bennybing@ieee.org to request for the URL. Constructive feedback on any aspect of the book is always welcomed.

The chapters in this book are organized as follows. Chapter 1 addresses several technical considerations related to wireless network operation which may have a direct impact on the feasibility of implementing multiple access protocols. In Chapter 2, the key issues for communication among distributed users who share and contend for a single broadcast channel are identified. The fundamental characteristics of multimedia traffic and the associated networking requirements are also reviewed. Chapter 3 defines the multiple access problem and focuses on the ideas and core concepts behind the major protocols that have been proposed over the years. It serves as a guide for those who wish to design new protocols that meet specific applications. This is followed a broad survey of existing multiple access strategies in Chapters 4 to 7. The strategies are categorized under fixed assignment, contention, reservation, and spread spectrum multiple access. Chapter 8 discusses several state-of-art broadband multiple access schemes that are applied in emerging high-speed wireless networks, including the IEEE 802.11, HiperLAN, HomeRF, Bluetooth, wireless ATM, satellite ATM, LMDS, and IMT-2000. This chapter provides a valuable resource for network designers who wish to understand the access procedures of these advanced networks. The complete design methodology of a generalized broadband wireless access protocol is described in Chapter 9. The performance is analyzed, simulated, and validated. The OPNET simulation models and protocol verification procedures are particularly useful for researchers working on multiple access protocol design. Finally, an extensive list of references allows readers to explore the latest research challenges in this highly fascinating subject. The references are separated according to the topics discussed in each individual chapter. The myriad of access protocols proposed over the last three decades is clear evidence of how important this field has become and will continue to be a fertile ground for research in the years ahead.

Writing a rapidly evolving technical subject is never an easy task. I am especially indebted to several individuals who have provided valuable feedback and personal insights which helped to refine certain aspects of the book. These include Professor Robert Gallager (Massachusetts Institute of Technology), Professor Leonard Kleinrock (University of California at Los Angeles), and Dr. Jouni Mikkonen (Nokia). At the same time, I would like to acknowledge the comments from the reviewers as well as the encouragement, patience, and help from Patricia Lincoln and Alex Greene from Kluwer Publishers. Special thanks also go to the production staff for an excellent job in producing the manuscript. Finally, I am grateful to my former thesis advisor, Professor Regu Subramanian at the Nanyang Technological University who first introduced me to this immense subject of multiple access communications.

Benny Bing
Maryland, USA

Chapter 1

OVERVIEW OF WIRELESS NETWORKS

The performance of a broadband wireless network is heavily dependent on the characteristics of the wireless channel such as signal fading, multipath distortion, limited bandwidth, high error rates, rapidly changing propagation conditions, mutual interference of signals, and the vulnerability to eavesdrop and unauthorized access. Moreover, the performance observed by each individual user in the network is different and is a function of its location as well as the location of other interacting users. In order to improve spectral efficiency and hence, the overall network capacity, wireless access techniques need to be closely integrated with various interference mitigation techniques including the use of smart antennas, multiuser detection, power control, channel state tracking, and coding.

1.1 SIGNAL COVERAGE

Perhaps one of the most important aspects of wireless system design is to ensure that sufficient levels of signal transmission are accessible from most of the intended service areas. Estimating signal coverage requires a good understanding of the communication channel, which comprises the antennas and the propagation medium. Usually, additional signal power is needed to maintain the desired link quality and to offset the amount of received signal power variation about its average level. These power variations can be broadly classified under small-scale or large-scale fading effects.

Small scale fades are dominated by multipath propagation, Doppler spread, and movement of surrounding objects. Large scale-fades are characterized by attenuation in the propagation medium and shadowing caused by obstructing objects. These effects are explained fully in the following sections.

1.2 PROPAGATION MECHANISMS

A transmitted signal diffuses as it travels across the wireless medium. As a result, a portion of the transmitted signal power arrives directly at the receiver while other portions arrive via reflection, diffraction, and scattering. Reflection occurs when the propagating signal impinges on an object that is large compared to the wavelength of the signal (e.g., buildings, walls, surface of the earth). When the path between the transmitter and receiver is obstructed by sharp, irregular objects, the propagating wave diffracts and bends around the obstacle even when a direct line-of-sight path does not exist. Finally, scattering takes place when the objects of obstruction are smaller than the wavelength of the propagating signal (e.g., people, foliage).

1.2.1 Multipath

When a signal takes multiple paths to reach the receiver, the received signal is split into different components (Figure 1.1), each with a different delay, amplitude, and phase. These components form different clusters and, depending on the phase of each component, interfere constructively and destructively at the receiving antenna, thereby producing a phenomenon called multipath fading.

Multipath fading represents the quick fluctuations in received power and is therefore commonly known as fast fading. It is affected by the location of the transmitter and receiver, as well as the movement around them. Such fading tends to be frequency-selective. Of considerable importance to wireless network designers is not only the depth but also the duration of the fades. Fortunately, it has been observed that the deeper the fade, the less frequently it occurs and the shorter the duration when it occurs. The severity of the fades increases as the distance between transceivers increase.

Since multipath propagation results in a variety of travel times, signal pulses are broadened as they travel through the channel. This limits the speed at which adjacent data pulses can be sent without overlap and hence, the maximum information rate a wireless system can operate. Thus, in addition to frequency-selective fading, a multipath channel also exhibits time dispersion. Time dispersion leads to intersymbol interference (ISI) while fading induces periods of low signal-to-noise ratio, both effects causing burst errors in digital transmission. Multpath fading is often classified as small-scale fading because the rapid changes in signal strength only occur over a small area or time interval.

The performance metric for a wireless system operating over a fading multipath channel is either the average probability of error or the probability of outage. The average probability of error is the average error rate for all possible locations in the coverage area. The probability of outage represents the probability of error below a certain signal threshold for all possible locations in the coverage area.

1.2.2 Delay Spread

Delay spread is caused by differences in the arrival time of a signal from the various paths when it propagates through a time-dispersive multipath channel. The delay spread is proportional to the length of the path, which is in turn affected by the size and architecture of the propagating environment as well as the location of the objects around the transmitter and receiver. A negative effect of delay spread is that it results in ISI. This causes data symbols (each representing one or more bits) to overlap in varying degrees at the receiver. Such overlap results in bit errors that increase as the symbol period approaches the delay spread. Clearly, the effect becomes worse at higher data rates and cannot be solved simply by increasing the power of the transmitted signal.

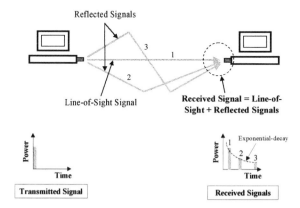

Figure 1.1: Multipath propagation

To avoid ISI, the data rate should not exceed the inverse of the delay spread. If high data rates cannot be avoided, then equalizers are typically required to combat the ISI problem. More sophisticated receivers employing adaptive equalizers can improve performance further by suppressing the multipath components. However, they must also rapidly obtain good estimates of the channel impulse response, which can be very difficult in time-varying wireless channels. Instead of an equalizer, wideband wireless systems based on spread spectrum transmission often use a simpler device called a correlator to counter the effects of multipath.

The root mean square (rms) delay spread is often used as a convenient measure to estimate the amount of ISI caused by a multipath wireless channel. The maximum achievable data rate depends primarily on the rms delay spread and not the shape of the delay spread function.

1.2.3 Coherence Bandwidth

A direct consequence of multipath propagation is that the received power of the composite signal varies according to the characteristics of the wireless channel in which the signal has traveled (Figure 1.2). The bandwidth of the fade (i.e., the range of frequencies that fade together) is called the coherence bandwidth. This bandwidth is inversely proportional to the delay spread.

Signals with bandwidth larger than the coherence bandwidth of the channel may make effective use of multipath by resolving many independent propagation paths. On the other hand, multipath interference can be avoided by keeping the data rate low, thereby reducing the signal bandwidth below the coherence bandwidth. Note that although a wideband receiver can resolve more paths than another receiver with a narrower bandwidth, the one with wider bandwidth receives less energy per resolvable path.

The delay spread caused by multipath is typically greater outdoors than indoors due to the wider coverage area. This gives rise to a higher coherence bandwidth in indoor environments. For example, indoor radio channels usually produce delay spreads of between 100 to 250 ns, which correspond to a coherence bandwidth of 4 to 10 MHz. In outdoor environments, the much larger delay spread of more than 1 μs implies a smaller coherence bandwidth of less than 1 MHz.

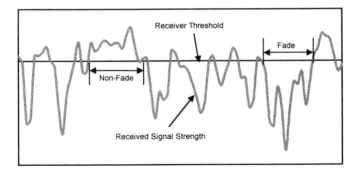

Figure 1.2: Signal fading characteristics

1.2.4 Doppler Spread

Doppler spread introduces random frequency/phase shifts at the receiver that can result in loss of synchronization. It is caused by the relative motion between the transmitter and receiver. It may also be due to the movement of reflecting objects that cause multipath fading. In an indoor environment for instance, the movement of people is the main cause of Doppler spread. A person walking at 10 km/hr can induce a maximum Doppler spread of ±22 Hz at 2.4 GHz.

The Doppler spread for indoor channels is highly dependent on the local environment, providing different shapes for different physical layouts. On the other hand, outdoor Doppler spreads consistently exhibit peaks at the limits of the maximum Doppler frequency. Typical values for Doppler spread are 10 to 250 Hz (suburban areas), 10 to 20 Hz (urban areas), and 10 to 100 Hz (office areas).

Just as coherence bandwidth is inversely related to the delay spread, coherence time is defined as the inverse of the Doppler spread. The relationship between coherence time and coherence bandwidth is illustrated in Figure 1.3. This relationship forms the basis for fading channel classification.

1.2.5 Shadow Fading

Besides multipath fading, physical obstructions (e.g., walls in indoor environments, buildings in outdoor environments) can cause large-scale shadow fading where the transmitted signal power is blocked and hence severely attenuated by the obstruction. The amount of shadow fading is dependent on the relative positions of the transmitter and receiver with respect to the large obstacles in the propagation environment. Unlike multipath fading (which is usually represented by a Rician or Rayleigh distribution), shadow fading is generally characterized by the probability density function of a lognormal distributed random variable.

1.2.6 Radio Propagation Modeling

An accurate characterization of the propagation mechanism is difficult since it is greatly influenced by a number of factors (e.g., antenna heights, terrain, topology). Radiowave propagation modeling is usually based on the statistics of the measured channel profiles (time and frequency domain modeling) or on the direct solution of electromagnetic propagation equations (Maxwell's equations and ray tracing).

The most popular models for indoor radio propagation are the time domain statistical models. In this case, the statistics of the channel parameters are collected from measurements in the propagation environment of interest at various locations between the transmitter and receiver. Ray tracing makes use of the fact that all objects of interest within the propagation environment are large compared to the wavelength of propagation, thus removing the need to solve Maxwell's equations. Its usefulness is ultimately dependent on the accuracy of the site-specific representation of the propagation medium.

1.2.7 Narrowband and Wideband Channel Models

Modeling the channel characteristics of narrowband and wideband signals is different. For narrowband signals, the emphasis is on the received power whereas for wideband communications, both the received power and multipath characteristics are equally important. A further distinction exists between models that describe signal strength as a function of distance as opposed to a function of time. The former is used to determined coverage areas and cochannel interference while the latter is use to determine bit error rates and outage probabilities.

1.3 SIGNAL ATTENUATION

All transmitted signals are attenuated (weakened) during propagation. Depending on the severity, the decay in signal strength can make the signal become unintelligible at the receiver. Analysis of radio propagation is very complex because the shortest direct path between the transmitter and receiver is often blocked by fixed and moving objects (e.g., hills, buildings, people). As a result, the received signal arrives by several reflected paths. The degree of attenuation depends to a great extent on the signal frequency. For example, lower frequencies tend to penetrate objects better while high frequency signals encounter greater attenuation.

1.3.1 Received Power Characteristics

For line-of-sight distances in the vicinity of the receiving antenna, signal attenuation is close to free space. This means that the received signal power decreases with the square of the distance between the transmitter and receiver. For instance, the signal strength at 2 m is a quarter of that at 1 m. At longer distances away from the receiving antenna, an increase in the attenuation exponent is common (Figure 1.4).

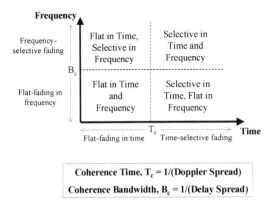

Figure 1.3: Coherence time and coherence bandwidth

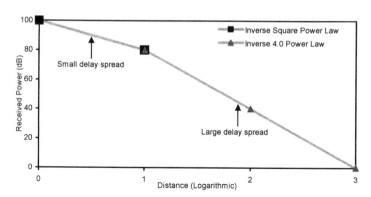

Figure 1.4: Received power characteristics

1.3.2 High Frequency Propagation

A high operating frequency implies an abundance of bandwidth. At high frequencies, oxygen absorption in the atmosphere leads to rapid fall off in signal strength and this facilitates frequency reuse in high-capacity, cellular-based wireless systems.

There are two oxygen absorption bands ranging from 51.4 GHz to 66 GHz (A band) and 105 GHz to 134 GHz (B band). There are also peaks in water vapor absorption at 22 GHz and 200 GHz. The oxygen absorption is lower in band B than band A while the water vapor attenuation is higher. These observations suggest that band A is more suitable for communications than band B. However, many technical challenges remain. In addition, cost considerations may prevent the use of such bands.

The Millimeter Wave Communications Working Group is defining how sharing of bandwidth can be achieved in 59 to 64 GHz band released by the FCC. Since the allocated spectrum is not licensed, large-scale frequency planning is avoided and ad-hoc networks are possible. On the downside, unlicensed spectrum allocation also leads to problems in controlling interference. Users must observe etiquette during transmission so that incompatible systems can co-exist. The key elements of etiquette are:

❑ Listening before transmitting;
❑ Limiting transmission time;
❑ Limiting transmit power.

1.4 CHANNEL CHARACTERISTICS

Many wireless systems show large performance gains when channel characteristics are known to the receiver. The extent to which these gains are achieved depends on the accuracy with which the receiver can estimate the channel parameters. In practical implementations, the fundamental characteristics of the channel are often affected by fading.

1.4.1 Gaussian Channel

The Gaussian channel is important for providing an upper bound on the system performance and accurately describes many physical time-varying channels. Often referred to as the additive white Gaussian noise (AWGN) channel, it is typically used to model the noise generated in the receiver when the transmission path is ideal (i.e., line-of-sight). The noise is assumed to have a constant power spectral density over the channel bandwidth and a Gaussian probability density function. When there is multipath fading and the user is stationary, the channel can still be approximated as Gaussian with the effects of fading represented as a path loss.

The Gaussian channel model has received considerable attention in the multiple access literature because of its importance in the case of single-transmitter and single-receiver channels. However, the Gaussian model may not be suitable when applied to multiple access channels where users transmit intermittently. In this case, the desired model is one where the number of active users on the network is a random variable and the Gaussian signals are conditioned on the fact that some user is transmitting [ABRA93]. In addition, Doppler spread, multipath fading, shadowing, and mutual interference from transmitting users make the channel far from Gaussian.

1.4.2 Rayleigh Channel

There are two kinds of channel fading namely, long-term (slow or log-normal) and short-term (fast or Rayleigh) fading. Long-term fading is characterized by the envelope of the fading signal which is related to the distance and the received power. Short term fading is primarily caused by reflections of a transmitted signal and refers to the rapid fluctuations of the received signal amplitude. In addition, second order statistics on fades exist and are usually functions of time (e.g., level crossing rates, average duration of fades, distribution of fade duration).

The received signal is a product of the long-term fading and short-term fading characteristics. The short-term fading signal is superimposed on an average value that varies slowly as the receiver moves (Figure 1.5).

1.4.3 Rician Channel

In some wireless channels, a dominant path (normally the direct line-of-sight path) exists between the transmitter and receiver, in addition to many scattered (non-direct) paths. This dominant path reduces the delay spread and may significantly decrease the fading depth, thus requiring much smaller fading margin in system design. The probability density function of the received signal envelope is said to be Rician.

1.5 FADING MITIGATION METHODS

Effective countermeasures against fading include antenna diversity, equalization, error control, and multicarrier transmission. These methods are discussed next.

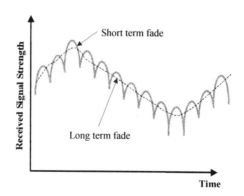

Figure 1.5: Short and long-term fading characteristics

1.5.1 Antenna Diversity

Under poor propagating conditions, the channel can be changed using antenna diversity. Macroscopic (spatial) diversity is important when the fading is slow (long-term). It is often the best approach because it allows receivers to choose independent channels without much overhead. Among the various forms of spatial diversity, selection combining is the simplest to implement while maximal ratio combining provides the largest signal-to-noise ratio improvement [ACAM87]. Microscopic diversity using two or more antennas or two frequencies at the same (collocated) antenna site can help reduce short-term fading. Polarization diversity can also be used to select the best channel at a particular location. Circular polarized directional antennas, when used in line-of-sight channels, can provide much lower delay spread than linear polarized antennas with similar directionality.

Active, narrow-beam smart antennas have been recognized as an effective means to increase system capacity and enlarge signal coverage area. Dielectric lens and patch arrays on soft substrates can be used to produce low-cost antennas that are nearly omni-directional in the azimuth direction but with narrow beams in elevation. Such antennas radiate energy only in a particular direction, providing gain along the intended direction and attenuation in the undesired directions. In doing so, a wireless coverage area can be broken down into smaller areas called sectors, each serviced by a directional antenna. This is equivalent to increasing the number of communication links which increases system capacity. Selection algorithms at the antennas assess the received signal strength and quality from different paths, giving the system a high probability of finding a path that is not corrupted by fading or interference, thereby mitigating undesirable multipath effects. Thus, interference and transmit power requirements are reduced. Adaptive antenna arrays are typically supported by connection-oriented pilot signals on both the up and down links. When employed by a base station, these antennas are able to track the locations of mobile users.

1.5.2 Equalization

Receivers can employ equalizers to compensate for ISI caused by a time-dispersive channel and regenerate the data symbols correctly. Equalizers essentially subtract delayed and attenuated images of the direct signal from the received signals. To do this, good estimates of the impulse response of the channel must be obtained, which implies that the channel must be linear. This response must also be frequently re-measured as the wireless channel can change rapidly in both time and space.

Equalizers require a training sequence at the start of each packet transmission to derive the frequency response (transfer characteristic) of the channel. They employ discrete-time filtering and generally fall into three broad classes namely,

❑   linear transversal equalizer (LTE);
❑   maximum likelihood sequence estimator (MLSE);
❑   decision feedback equalizer (DFE).

An equalizer can be made adaptive by employing an adaptive filter at the receiver whose frequency response adapts to the inverse of the frequency response of the channel.

LTEs operate by viewing channel distortion as a filtering process and then attempting to construct an inverse filter in the receiver. Although such equalizers are the simplest to implement, their performance is limited because the inverse filter of the wireless channel may not exist in all cases or may cause the signal to be restored to its original form at the expense of boosting the received noise to a very high level.

The MLSE (also known as Viterbi equalizers) operates by testing hypotheses of the transmitted data sequence, combined with knowledge about the channel's impulse response, against the signal which was received. The signal that gives the closest match is chosen. Although this process achieves the lowest error probability, it is quite complex and is not feasible at high data rates.

LTEs are not effective on frequency-selective multipath fading channels. Instead, DFEs are typically used on these channels. Like spread spectrum transmission, a DFE can isolate the arriving paths and take advantage of them as a source of implicit diversity to improve performance. A DFE feeds past decisions through a feedback filter and subtracts the results from the input to the decision device. This non-linear processing results in less noise enhancement than linear equalization. Considerably simpler than the MLSE but with much better performance than a LTE, a DFE also exploit decisions about the received data to cancel ISI from the received signal. A

DFE is easier to implement than a LTE since the input to its feedback filter are data symbols. Hence, convolution can be implemented based solely on additions and no multiplications are required. However, the level of processing power required for such a device is beyond the capabilities of a general signal processor and an application-specific IC (ASIC) is necessary.

Sequence detection can offer performance improvements over traditional equalization when the channel response is known and time-invariant. It can be made adaptive for unknown and/or time varying channels with the use of an adaptive channel estimator.

1.5.3 Error Control

Two general types of error control are typically employed on wireless channels. Forward error correction (FEC) techniques use error control codes (e.g., convolutional coding, block coding) that can detect with high probability, the error location. In automatic repeat request (ARQ), the receiver employs error detection codes to detect errors in the received packets and then request the transmitter to resend any error packet.

FEC tend to degrade throughput and may require long interleaving which increases transmission delay. On the other hand, ARQ schemes are simpler and more flexible to implement but suffer from variable delay.

1.5.4 Multicarrier Transmission

At high data rates, the computation complexity for an equalizer increases. In addition, the overhead for channel estimation increases when the channel is time-varying. Multicarrier (multitone) systems are parallel transmission schemes that compensate for the multipath delay spread without the need for equalization. The principle behind multicarrier schemes is that since ISI due to multipath is only a problem for very high signaling rates (typically above 1 Mbit/s), one can reduce the signaling rate until there is no degradation in performance due to ISI. The overall frequency band is divided into a number of subchannels, each modulated on a separate subcarrier at a much lower symbol rate than a single carrier scheme using the entire bandwidth. Since each subchannel is narrow enough to cause only flat (non frequency-selective) fading, this makes a multicarrier system less susceptible to ISI. By adding a small guard interval, such interference can be completely eliminated. To obtain the required throughput, one transmits a number of these relatively low bit rate signals simultaneously on adjacent frequency channels.

Multicarrier modulation is spectrally efficient because the subcarriers are packed close together. It also allows considerable flexibility in the choice of different modulation methods. On the negative side, OFDM is more sensitive to frequency offset and timing mismatch than single-carrier systems. Like multilevel schemes, a major drawback of multicarrier systems is the high peak-to-average power ratio. For N carriers, if the peak power is limited, then the average power that can be transmitted is reduced by $1/N$. The need to amplify a set of frequency carriers simultaneously implies the presence of signal envelope variation. This requires the use of a highly linear (and inefficient) amplifier, which leads to high power consumption. Such an amplifier is also needed to suppress intermodulation interference caused by the transmission of multiple frequency carriers.

1.5.5 Orthogonal Frequency Division Multiplexing

The concept of Orthogonal Frequency Division Multiplexing (OFDM) has been employed in discrete multitone (DMT) systems (e.g., ADSL, European DAB). OFDM is a special form of multicarrier modulation and hence, inherits all the advantages (and disadvantages) of multicarrier modulation. However, an OFDM system uses less bandwidth than an equivalent multicarrier system because the frequency subchannels are overlapped. This is achieved using orthogonal signals in the subchannels. OFDM offers frequency diversity which can be exploited by proper coding to combat frequency-selective fading.

OFDM signals can be generated and decoded using Fast Fourier Transform (FFT), which can be adapted to different data rates and different link conditions. Since both transmitter and receiver modulation can be achieved using FFT, this allows efficient digital signal processing implementation and significantly reduces the amount of required hardware compared to conventional Frequency Division Multiple Access (FDMA) systems. Of considerable importance in the design of OFDM systems is frequency synchronization and power amplifier backoff in the receiver. Moreover, the number of subcarriers has to be chosen in an appropriate way.

1.5.6 Wideband Systems

In wideband systems (e.g., spread spectrum systems), the transmission bandwidth of a signal is much greater compared to the channel bandwidth. As a result, multipath fading affects only a small portion of the signal bandwidth.

Like narrowband systems, multipath, shadow fading and interference impose limits on the coverage region and on the number of users in wideband systems. For example, if a 100 Mchips/s chip rate is selected for a spread spectrum system, then multipath components must be separated by at least 10 ns in order for these components to be resolved. Hence, the appropriate area of coverage should have a propagation delay of greater than 10 ns or equivalently, a minimum size of 3×10^8 m/s $\times$ 10 ns or 3 m.

In general, a bigger coverage area is preferred for spread spectrum communications since this results in more resolvable multipaths. However, the opposite is true for narrowband systems. Furthermore, polarization antenna diversity becomes more effective since there is significant coupling between the vertical and horizontal polarization directions when coverage area is wide.

1.6 INTERFERENCE

A key feature of mobile cellular systems involving TDMA is that the same set of frequency channels is reused after a respectable distance (Figure 1.6). A frequency channel is often assigned to each user in a specific radio coverage area. When the user crosses a radio cell boundary, the assigned frequency channel changes. Such spatial reutilization of bandwidth achieves a higher overall spectrum utilization by allowing multiple transmissions to take place simultaneously in different locations without causing excessive interference A radio cell operating on one frequency channel is protected from the mutual interference arising in reused frequency channels by a surrounding ring of adjacent cells operating at other frequency channels. Ultimately, it is this mutual interference that limits the performance of such cellular systems. In spread spectrum CDMA systems, the concept of frequency reuse translates to code reuse where the same code is assigned to two users if they are far apart from one another.

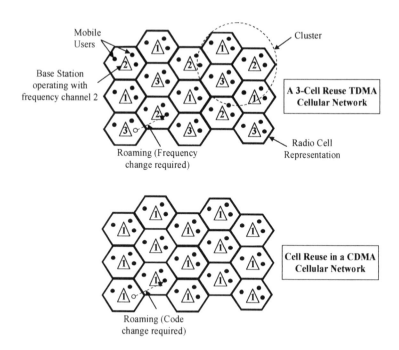

Figure 1.6: Spatial utilization in cellular systems

1.6.1 Cochannel Interference

A common type of interference encountered in systems that employ frequency reuse is cochannel interference (CCI). TDMA users experience CCI interference from time-overlapping slots that originate from users transmitting using the same reuse frequencies in different radio cells. Such intercell interference can be minimized through proper frequency planning. CCI is a major problem encountered in direct sequence spread spectrum (DSSS) systems. In this case, CCI is a combination of multiaccess interference (MAI) due to transmission from multiple users as well as intersymbol interference (ISI) arising from multipath. The overall interference, which increases as the number of transmitting users increases, severely limits the capacity of DSSS systems.

It is important to realize that the performance of a spread spectrum receiver can vary widely even though the number of transmitting users is fixed. If the interfering users happen to be close to the receiver, MAI can be very large. Conversely, if the interfering users are far from the receiver, then MAI can be very small. Thus, the geographic location of the receiver also influences the amount of interference experienced by the user and this changes as users move to different locations.

1.6.2 Mitigation Techniques

Several CCI mitigation techniques commonly employed by TDMA and FDMA systems are listed in Table 1.1. Most of these methods will be covered in the later chapters.

1.7 MODULATION

The maximum data rate is not only bounded by the multipath characteristics of the channel but also the modulation technique used. With a bandwidth-efficient modulation technique, a greater number of bits can be transmitted in each symbol interval, resulting in higher data rates (Figure 1.7). The main considerations for selecting a modulation scheme include bandwidth (spectral) and power efficiency, resistance against multipath, cost/complexity of implementation, constant envelope, and out-of-band radiation.

1.7.1 Linear versus Constant Envelope

Constant envelope (e.g., GMSK) and linear modulation (e.g., QAM) are commonly employed in high-speed wireless networks. Constant envelope modulation involves only the phase whereas non-constant envelope modulation requires both the phase and the amplitude to be modulated.

Table 1.1: Interference mitigation techniques

TDMA	CDMA
Control reuse patterns	Control spreading gain
Implement loose power control	Implement tight power control
Implement guard times	Implement antenna arrays
Implement error control codes	
Implement antenna arrays	

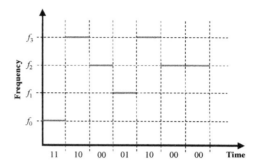

Figure 1.7: Multilevel modulation

Constant envelope modulation has gains in power efficiency since it allows more efficient Class C power amplifiers to be used. On the other hand, linear modulation is spectrally more efficient than constant envelope modulation but requires the use of linear power amplifiers. The concern for linearity is primarily due to stringent restrictions on intermodulation products and out-of-band power emission requirements.

There is considerable debate as to whether multilevel modulation provides any real gain in frequency reuse situations since an increase in carrier-to-interference ratio (required in multilevel modulation) results in an increase in frequency reuse distance, which more than offsets the modulation improvement.

Both π/4-QPSK (1.8 bit/s/Hz) and GMSK (1.6 bit/s/Hz) are bandwidth-efficient modulation schemes that are widely employed in wireless systems. Since GSMK provides smoothing of phase transitions at symbol boundaries, it has superior adjacent channel interference performance. GMSK, however, exhibits high error floors and needs extra care in the design of its demodulator to ensure reliable carrier and timing recovery in multipath environments. In contrast, a reliable carrier can be more easily extracted with π/4-QPSK modulation using differential detection.

1.7.2 Coherent versus Non-Coherent Detection

There are two basic techniques of detecting data symbols from the demodulated signals namely, coherent (synchronous) and non-coherent (envelope) detection. In coherent detection, the receiver first processes the signal with a local carrier of the same frequency and phase, and then cross-correlate with other replicated signals received before performing a match (within a predefined threshold) to make a decision. On the other hand, non-coherent detection does not depend on a phase reference and is therefore less complex but exhibits worse performance when compared with coherent methods. Some common digital modulation methods with coherent and non-coherent detection are listed in Table 1.2. In a multipath environment, carrier synchronization is not easy and hence, non-coherent detection is preferred. Note that although no absolute phase reference is needed for the demodulation of differential phase modulation, the receiver is sensitive to carrier frequency offset.

1.7.3 Multicarrier Modulation

Just as WDM is boosting the bandwidth of optical fiber systems many times through the simultaneous use of multiple wavelength channels, multicarrier modulation relies on sending several low data rates in parallel that combine to the desired high rate transmission. It is becoming an increasing popular alternative to equalization.

The principle behind multicarrier schemes is that since ISI due to multipath is only a problem for very high data rates (typically above 1 Mbit/s), the data rate can be reduced until there is no degradation in performance due to ISI. To obtain the required data throughput, a number of these relatively low bit rate signals are transmitted simultaneously on adjacent frequency channels. The main drawback with this approach is that a highly linear (and inefficient) amplifier is required to suppress intermodulation interference.

Table 1.2: Coherent and non-coherent detection

Coherent	Non-Coherent
Amplitude shift keying	Amplitude shift keying
Frequency shift keying	Frequency shift keying
Phase shift keying	Differential phase shift keying
Continuous phase modulation	Continuous phase modulation

1.8 SIGNAL DUPLEXING TECHNIQUES

The duplexing scheme is usually described along with a particular multiple access scheme. This section provides an assessment of the relative strengths and weaknesses of Frequency Division Duplex (FDD) and Time Division Duplex (TDD).

1.8.1 Spectrum Considerations

The amount of spectrum required for both FDD and TDD is similar. The difference lies in the fact that FDD employs two bands of spectrum separated by a certain minimum bandwidth (guard band), while TDD requires only one band of frequencies. TDD has an advantage here since it may be easier to find a single band of unassigned frequencies than to find two bands of unassigned frequencies separated by the required bandwidth.

1.8.2 Radio Design Considerations

Since FDD employs different frequencies in the two directions of transmission, diversity antennas have to be employed at both the base station and the user. TDD, however, benefits from antenna diversity without the need for multiple antennas. This is attributed to the fact that for a given frequency, the attenuation in a radio channel is typically reciprocal. For example, a TDD system can have the base station select the best signal from its antenna when receiving a specific signal from a user. When the base station next transmits to that user, it employs the same antenna. Note that while the channel is reciprocal in attenuation, this may not apply to the carrier to interference ratio. Thus, dual antennas may still be needed for a TDD system.

Another antenna-related design consideration when selecting a duplex scheme is whether a duplexer is required. A duplexer adds weight and cost to a radio transceiver, and can place a limit on the minimum size of a mobile device. This is because dual channels imply more complex receivers than a single channel system where all receivers operate at the same frequency. TDD is a burst mode transmission scheme. During the transmission, the receiver is deactivated. Thus, TDD systems are capable of providing bidirectional antenna diversity gain without employing duplexers.

However, in terms of equipment utilization, a TDD transceiver effectively remains idle half of the time.

Although duplexers are used on most FDD systems, they are not required in FDD systems employing time division multiple access (TDMA) since the transmit and receive time slots occur at different times. A simple RF switch performs the function of the duplexer, but is less complex, smaller in size, and cheaper. Such a switch connects the antenna to the transmitter when a transmit burst is required and to the receiver for the incoming signal.

1.8.3 Implementation Considerations

Since FDD uses different frequencies for each direction of transmission, interference is not possible even if the timing on the two frequencies is not synchronized. However, on each TDD link, precise synchronization is required. Otherwise, overlapping transmit and receive bursts will result in a reduction of overall system capacity [DUET93].

Another radio implementation consideration when selecting a duplex scheme is the amount of transmission delay incurred. In the case of TDD, each transmit burst is followed by a receive burst and this separation produces the delay. The introduction of such delay is inherent in any time division techniques (e.g., TDMA) that employ a fixed frame structure.

Finally, an important advantage of TDD over FDD is the ability to support asymmetric traffic loads since channel transmission time in TDD can be apportioned flexibly between the up and down links. Internet access is primarily asymmetric in nature.

1.9 MOBILITY AND HANDOFF

Tracking a mobile user in a wireless environment creates additional complexity since it requires rules for handoff and roaming. In addition, one must also cope with rapid fluctuations in received signal power and traffic load patterns that changes with user location and time.

Handoff refers to the process of changing communication channels so that uninterrupted service can be maintained when users move across radio cell boundaries. Clearly, a handoff between two radio cells will change the number of active connections in each cell and this in turn affects the traffic conditions and interference level in the cells.

An effective handoff scheme must take into consideration three key factors:

❑   propagation conditions;
❑   traffic load;
❑   switching and processing requirements.

It requires a neighboring base station to have a free channel with good signal quality and that the switchover be completed before any significant deterioration to the existing link occurs. The handoff process essentially comprises two stages namely,

❑   Link quality evaluation and handoff initiation;
❑   Allocation of bandwidth resources.

1.9.1 Intercell versus Intracell Handoff

When the user crosses two adjacent radio cells, an intercell handoff is required so that an acceptable link quality for the connection can be maintained. Sometimes an intracell handoff is desirable when the link with the serving base station is affected by excessive interference, while another link with the same base station can provide better quality.

1.9.2 Mobile-Initiated versus Network-Initiated Handoff

Network controlled handoff (NCHO) has been widely used in first generation analog cellular systems (e.g., AMPS). With NCHO, the link quality is monitored only by base stations. The handoff decision is made under the centralized control of a mobile telephone switching office (MTSO). Typically, NCHO algorithms only support intercell handoffs, have handoff network delays of in the order of several seconds, and have relatively infrequent updates of the link quality from the alternate base stations.

Mobile assisted handoff (MAHO) is a decentralized strategy used in several digital cordless telephone systems (e.g., DECT). The link quality is measured by both the serving base station and the user. In TDMA systems, the user measures the signal strength during the intervals when it is not allocated a time slot. Link quality measurements of alternate base stations are done by the user. Link measurements at the serving base station are relayed to the user. The handoff decision is made by the user. The MAHO algorithm has the lowest delay (about 100 ms) and exhibits high reliability. Both intracell and intercell handoffs are supported.

1.9.3 Forward versus Backward Handoff

Handoff algorithms also differ in the way a connection is transferred to a new link. Backward handoff algorithms initiate the handoff process through the currently serving base station. No access to the new link is made until resources have been allocated. The advantage of backward algorithms is that the signaling information is transmitted through an existing radio link and therefore, the establishment of a new signaling channel is not required during the initial stages of the handoff process. The disadvantage is that the algorithm may fail in conditions where the link quality with the serving base station is rapidly deteriorating. This type of handoff is used in most TDMA cellular systems (e.g., GSM).

Forward handoff algorithms activate the handoff process via a channel with the target (alternate) base station without relying on the current base station. Although the handoff process is faster, this is achieved at the expense of a reduction in handoff reliability. Forward handoffs are popular in digital cordless telephone systems (e.g., DECT).

1.9.4 Hard versus Soft Handoff

Handoffs can also be hard or soft. Hard handoffs release the radio link with the current base station at the same time when the new link with the target base station is established. Such handoffs are used in most TDMA cellular systems (e.g., IS-54, PDC, GSM). Soft handoffs maintain radio links with at least two base stations. A link is dropped only when the signal level falls below a certain threshold. Soft handoff is used in CDMA cellular systems (e.g., IS-95).

1.10 CHANNEL ASSIGNMENT STRATEGIES

The assignment of frequency channels to radio cells and to users is a fundamental operation of a mobile communication system. A classification of channel assignment schemes is shown in Figure 1.8 [TEKI91]. In basic fixed channel assignment, channels that are assigned to cells for a relatively long period of time. This method obviously results in poor utilization of the available bandwidth when traffic patterns change over time. Dynamic channel assignment represents the opposite extreme where channels are assigned to radio cells only when required. Such techniques can adapt to traffic load changes in real time but suffer from increased network management overhead.

Between the extremes of fixed and dynamic channel assignment lie flexible channel assignment, channel borrowing, and hybrid channel assignment. Flexible channel assignment is basically fixed assignment altered regularly according to predicted changes in the traffic load. Borrowing strategies can also be considered to be a variant of fixed channel assignment where channels not in use in their allocated cell can be temporarily transferred to congested cells on a connection-by-connection basis. Hybrid assignment allocates a fraction of the channels according to fixed assignment and the rest according to dynamic assignment.

1.11 SYNCHRONIZATION

An assumption common to all slotted multiple access protocols (e.g., slotted ALOHA, TDMA) is that each user is synchronized to a time reference, allowing transmissions by each user to arrive at the intended receiver at a time agreed upon by all users. For accuracy, this requires two items of information at each user:

❑   a common clock reference;
❑   forward plus reverse propagation times.

The latter allows each user to compensate for differences between its distance to the receiver and the distances of other users, a task that becomes increasingly difficult with high data rate channels.

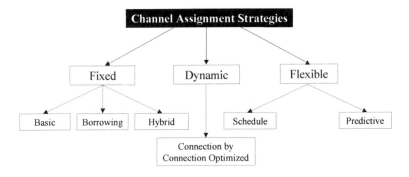

Figure 1.8: Channel assignment classification

1.12 POWER MANAGEMENT

Minimizing transmitted signal energy is an important consideration in power-limited mobile communication systems. This issue can be addressed at both the physical layer and higher network layers. Many approaches to reducing transmission energy can lead to interactions between network layers which are not present in wired, point-to-point networks. For example, the transmission rate and energy at the physical layer can be viewed as adjustable parameters by the higher layers. This provides new degrees of freedom for higher layer issues such as buffer control and routing. How these parameters can be adjusted depends on physical layer issues (e.g., coding, modulation) as well as the multiple access technique.

1.13 SPECTRUM ALLOCATION

Spectrum allocation plays an important role in wireless networking. However, it is difficult to locate radio spectrum that is available worldwide. This is because spectrum allocation is controlled by multiple international regulatory bodies (e.g., FCC in the United States, MKK in Japan, CEPT in Europe) as shown in Figure 1.9.

The 2.4 GHz Industrial, Scientific and Medical (ISM) band has been adopted by the IEEE 802.11 wireless LAN standard [BING00a]. Another band that has attracted considerable interest lately is the Unlicensed National Information Infrastructure (U − NII) frequency band. The large amount of radio spectrum (300 MHz) allocated by the FCC in January 1997 enables the provision of high-speed Internet and multimedia services. The 5 GHz U – NII band has been targeted by the wireless ATM community in its efforts to produce a global standard. The IEEE 802.11 has recently produced a high-speed wireless LAN standard based on the U-NII band. Finally, a total bandwidth of 230 MHz in the 1885 – 2025 and 2110 – 2200 MHz bands has been adopted by IMT-2000.

SUMMARY

Among the various forms of radio signal degradations, multipath fading assumes a high degree of importance. At low data rates, multipath causes Rayleigh fading. At high data rates (i.e., when the delay spread becomes comparable to the symbol interval), the received signals become indistinguishable, giving rise ISI. Such fading can be countered effectively using diversity techniques, in which two or more independent channels are somehow combined. The motivation here is that only one of the channels is likely to suffer a fade at any instant of time.

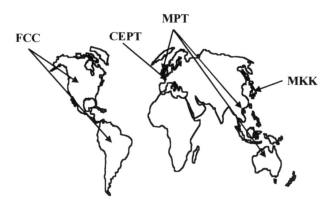

CEPT – European Conference of Postal and Telecommunications (Europe)
FCC – Federal Communications Commission (US)
MKK – Radio Equipment Inspection and Certification Institute (Japan)
MPT – Ministry of Post and Telecommunications

Figure 1.9: Worldwide spectrum regulatory bodies

BIBLIOGRAPHY

[ABRA93]   Abramson, N., *Multiple Access Communications*, IEEE Press, 1993.

[ACAM87]   Acampora, A. and Winters, J., "System Applications for Wireless Indoor Communications", *IEEE Communications Magazine*, Vol. 25, No. 8, August 1987, pp. 11 – 20.

[BERT00]   Bertoni, H., *Radio Propagation for Modern Wireless Systems*, Prentice Hall, 2000.

[BING90]   Bingham, J., "Multicarrier Modulation for Data Transmission: An Idea Whose Time Has Come", *IEEE Communications Magazine*, May 1990, pp. 5 – 14.

[BING00a] Bing, B., *High-Speed Wireless ATM and LANs*, Artech House, 2000.

[BING00b] Bing, B., "Guest Editorial: Multiple Access for Broadband Wireless Networks", *IEEE Communications Magazine*, July 2000.

[CHIA92]   Chia, S., "The Universal Mobile Telecommunication System", *IEEE Communications Magazine*, Vol. 30, No. 12, December 1992, pp. 54 – 62.

[DUET93]   Duet, D., Kiang, J. and Wolter, D., "An Assessment of Alternative Wireless Access Technologies for PCS Applications", *IEEE Journal on Selected Areas in Communications*, August 1993, pp. 861 – 869.

[EVER94]   Everitt, D., "Traffic Engineering of the Radio Interface for Cellular Mobile Networks", *Proceedings of the IEEE*, Vol. 82, No. 9, September 1994, pp. 1371 – 1382.

[GARG96]   Garg, V. and Wilkes, J., *Wireless and Personal Communications Systems*, Prentice Hall, 1996.

[GIBS96]   Gibson, J., ed., *The Mobile Communications Handbook*, CRC Press, 1997.

[GOOD98]   Goodman, D., *Wireless Personal Communications*, Addison Wesley, 1995.

[HASH93]   Hashemi, H., "The Indoor Radio Propagation Channel", *Proceedings of the IEEE*, Vol. 81, No. 7, July 1993, pp. 941 – 967.

[IEEE87a] "Special Issue on Mobile and Portable Communications Services", *IEEE Communications Magazine*, Vol. 25, No. 6, June 1987.

[IEEE87b] "Special Issue on Portable and Mobile Communications", *IEEE Journal on Selected Areas in Communications*, Vol. SAC-5, No. 5., June 1987.

[JACO78]   Jacobs, I., Binder, R. and Hoversten, E., "General Purpose
            Packet Satellite Networks", *Proceedings of the IEEE*, Vol. 66, No.
            11, November 1978, pp. 1448 – 1467.
[KATZ96]   Katzela, I. And Naghshineh, M., "Channel Assignment
            Schemes for Cellular Mobile Telecommunication Systems: A
            Comprehensive Survey", *IEEE Personal Communications*, Vol. 3.,
            No. 3, June 1996, pp. 10 – 31.
[LEE94]   Lee, E. and Messerschmitt, D., *Digital Communications*, Kluwer
            Academic Publishers, 1994.
[LI95]   Li, V. and Qiu, X., "Personal Communications Systems",
            Proceedings of the IEEE, September 1995, pp. 1208 – 1243.
[PAHL91]   Pahlavan, K., "Wideband Frequency and Time Domain
            Models for the Indoor Radio Channel", *Proceedings of IEEE
            GLOBECOM*, 1991, pp. 1135 – 1140.
[PAHL95]   Pahlavan, K. and Levesque, A., *Wireless Information Networks*,
            John Wiley, 1995.
POLL96]   Pollini, G., "Trends in Handover Design", *IEEE
            Communications Magazine*, Vol. 34, No. 3, March 1996, pp. 82 –
            90.
[RAPP95]   Rappaport, T., *Cellular Radio and Personal Communications*, IEEE
            Press, 1995.
[RAPP96a]   Rappaport, T., *Cellular Radio and Personal Communications:
            Advanced Selected Readings*, IEEE Press, 1996.
[RAPP96b]   Rappaport, T., *Wireless Communications*, Prentice Hall, 1996.
[SALE87]   Saleh, A., and Valenzula, R., "A Statistical Model for Indoor
            Multipath Propagation", *IEEE Journal on Selected Areas in
            Communications*, Vol. SAC-5, No. 2, February 1987, pp. 128 –
            137.
[SENA95]   Senarath, G. and Everitt, D., "Performance of Handover
            Priority and Queueing Systems under Different Handover
            Request Strategies, *Proceedings of the 45th IEEE Vehicular
            Technology Conference*, 1995, pp. 897 – 901.
[SKLA88]   Sklar, B., *Digital Communications*, Prentice Hall, 1988.
[STUB98]   Stuber, G., *Principles of Mobile Communications*, Kluwer Academic
            Press, 1998.
[TEKI91]   Tekinay, S. and Jabbari, B., "Handover and Channel
            Assignment in Mobile Cellular Networks", *IEEE
            Communications Magazine*, Vol. 29, No. 11, November 1991, pp.
            42 – 46.

[WOER95] Woerner, B., Rappaport, T. and Reed, J., *Wireless Personal Communications: Research Developments*, Kluwer Academic Publishers, 1995.

[VALE96] Valenzuela, R., "Antennas and Propagation for Wireless Communications", *Proceedings of the 46th IEEE Vehicular Technology Conference*, 1996, pp. 1 – 5.

[WINT94] Winters, J., Salz, J. and Gitlin, R., "The Impact of Antenna Diversity on the Capacity of Wireless Communication Systems", *IEEE Transactions on Communications*, Vol. 42, No. 2/3/4, February/March/April 1994, pp. 1740 – 1750.

Chapter 2

WIRELESS ACCESS PROTOCOL DESIGN

The basic element of a wireless access protocol is to allow a large group of uncoordinated users to share the same communications resource in a mutually cooperative and efficient manner. The choice of an efficient access protocol depends strongly on both the nature of the traffic and the performance demanded by the users. In addition, since the multiple access function is directly responsible for scheduling access to bandwidth resources, it is closely related with the resource management function.

2.1 TRAFFIC SOURCE CHARACTERIZATION

Traffic sources can be characterized by two stochastic (random) processes:

❑   message generation process;
❑   message length distribution process.

Two very important traffic sources are periodic (inelastic) and bursty (elastic) traffic. Since multimedia applications generate both periodic and bursty traffic, the need to support these traffic types with a guaranteed performance has to be addressed.

2.1.1 Periodic Traffic

Periodic traffic generates a stream of messages with very small interarrival variance between messages. Such traffic cannot adapt to delay variations (hence the alternative name "inelastic"). In addition, periodic traffic is delay sensitive because a maximum delay is allowable after which the information generated is no longer useful. Thus, periodic traffic requires preferential treatment in terms of the maximum end-to-end delay and the delay variation (jitter). All real-time applications that require time-based information to be presented at specific instants (e.g., voice, video) have periodic traffic patterns.

Digital video represents an important class of periodic traffic because it generates vast amounts of information, far exceeding the capacity of current networks or mass storage devices. For example, the digital representation of NTSC video with a spatial resolution of 800 × 600 pixels per frame, at 24 bits per pixel, and a temporal resolution of 30 frames/s results in a transmission rate of 345.6 Mbit/s. Without compression, multimedia networks involving video are unlikely to exist. Thus, the efficient representation of video signals (also called source coding) has been the subject of considerable research over the past 25 years.

Efficient video coding schemes (e.g., MPEG) attempt to maintain the same quality for all images at the expense of producing variable bit rate (VBR) output. In circuit-switched networks, coding schemes must reduce resolution during complex scenes to achieve the constant bit rate (CBR) requirement. As a result, constant image quality cannot be maintained for all scenes. In packet-switched networks, where bandwidth can be allocated on demand and statistical multiplexing is used, VBR coding schemes can in principle be used without sacrificing bandwidth utilization efficiency. Smoothing is also typically performed on video traffic at the source. This helps to minimize the loss and delay, potentially increasing the amount of traffic that can be admitted into the network.

Several digital video standards that form the basis of many broadband services have been developed. Popular standards include the ITU-T H.261 standard for teleconferencing and the ISO/IEC MPEG family of standards. These standards are all based on the same general architecture namely, motion-compensated temporal coding coupled with block discrete cosine transform (DCT) spatial coding. They employ hierarchical video coding to support interoperability between different services and to allow receivers to operate with different capabilities (e.g., receivers with different spatial resolutions). For real-time transmission of broadcast video services over wireless networks, the intraframe to interframe ratio and the quantizer scale are two key parameters that can be used to control the video source.

2.1.2 Bursty Traffic

Bursty traffic is characterized by messages of arbitrary lengths generated at random time instants, typically separated by interarrival intervals (idle time) of random duration. Moreover, these interarrival intervals are usually of durations much longer than the periods of transmission, resulting in a high peak-to-average data rate ratio. The unpredictability of bursty traffic, especially the instants at which messages are generated, is the main issue that has to be addressed. In addition, the average message length is also an important parameter because it is possible for a traffic source to generate very long messages and still be considered bursty.

The bursty nature of a traffic source is more than just randomness in message generation time and length. The user-specified message delay constraints to be met are actually the single most important factor in determining if the traffic source behaves in a bursty manner [LAM83]. Based on two user-dependent metrics namely, the message arrival rate (λ) and the average message delay constraint (D), the burstiness of the traffic source (B) is defined to be the ratio of the average interarrival time between messages over the average message delay constraint:

$$B = \frac{1}{\lambda D}$$

(2.1)

B gives the upper bound on the duty cycle of the traffic source. A traffic source with small B ($\ll 1$) is said to be bursty. Clearly, the use of dedicated, circuit-switched communication links for bursty traffic results in poor channel utilization since enough capacity must be assigned to meet the peak (and not average) demand.

Unlike periodic traffic, bursty traffic can cope with wide changes in delay or throughput. Some examples include:

- file transfer (sensitive to throughput);
- e-mail (insensitive to delay);
- network management traffic (sensitive to delay during heavy network congestion when important management information need to be sent);
- Web browsing (sensitive to delay).

2.2 CHARACTERIZING APPLICATIONS

A key requirement for networks carrying multimedia traffic is that multiple sources, each requiring a different QoS, must be allocated the appropriate bandwidth. To achieve this objective, these statistically multiplexed traffic sources must be properly characterized. An application can be characterized by its information type and delivery requirements. A multimedia application can send any combination of real-time and nonreal-time transfers of time-based and nontime-based information.

2.2.1 Information Types

In general, the information to be communicated can be divided into time-based or nontime-based. Time-based information must be presented at specific instants to convey its meaning. Typical time-based information includes packetized video and audio while still-images, graphics, and text are classified under nontime-based information.

2.2.2 Delivery Requirements

Applications can also be categorized according to information delivery requirements namely, real-time or nonreal-time. A real-time application requires sufficient bandwidth for information to be delivered quickly for receiver playback. Nonreal-time applications require sufficient storage in order for information received to be stored for later use. An example of a real-time application is a voice conversation whereas sending an electronic mail is a nonreal-time application. Thus, communicating parties for real time applications participate at the same time while for a nonreal-time application, they participate at different times.

It is important to distinguish between the delivery requirement (which can be real-time or nonreal-time) of an application from the intrinsic time dependency of its information content (which can be time-based or nontime-based). Video conferencing and image browsing are some examples of real-time applications while downloading digitized movies and electronic mail belong to nonreal-time applications.

2.2.3 Symmetry of Connection

A communicating application usually involves two-way information transfer. Such bidirectional connections can be classified as either symmetric or asymmetric connections. The asymmetric nature of data traffic flow usually means that the volume and speed of data transfer in one direction is significantly different from the other. A voice connection over the Internet is an example of symmetric connection while Web browsing is generally considered an asymmetric connection that involves sending control information in one direction and data transfer in the reverse direction.

2.2.4 Communication Requirements

The communications requirements of an application fall under three categories namely,

- Bandwidth;
- Delay;
- Error.

Of the three, the bandwidth requirement is most critical because it immediately determines whether an application can be supported or not.

The bandwidth requirement of an application depends on its information type and delivery requirements. For real time applications that generate time-based information, the bandwidth requirement (which may be constant or variable) is simply the amount of information generated by the application per unit time. For other applications namely, either nonreal-time applications or applications sending nontime-based information, the bandwidth is a function of the response time requirement and the amount of information to be transmitted.

2.2.5 Broadband Services

The bandwidth requirements of common communicative and distributive broadband services are listed in Table 2.1. Note the symmetry and asymmetry in the uplink and downlink bandwidth requirements.

Table 2.1: Uplink and downlink bandwidth requirements for broadband services

Broadband Service	Downlink Bandwidth	Uplink Bandwidth
Broadcast video Enhanced pay per view	1.5 – 6 Mbit/s	Not applicable
Interactive video Video-on-demand Interactive games Information retrieval	64 Kbit/s – 6 Mbit/s	9.6 – 64 Kbit/s
Internet access	14.4 Kbit/s – 10 Mbit/s	14.4 Kbit/s – 128 Kbit/s
Video conferencing Video telephony	9.6 Kbit/s – 2 Mbit/s	9.6 Kbit/s – 2 Mbit/s

2.3 RESOURCE SHARING

Resource sharing procedures can be very complex because the demands have arbitrary arrival times and holding times (time-usage of resources). This is particularly true for multimedia traffic where the information to be transmitted can come from a rich variety of sources (e.g., voice, video, image, text) with different source characteristics (e.g., data rates, activity ratio, burstiness), and different quality of service (QoS) requirements (e.g., bit error rate, delay constraints). The problem involves a tradeoff between providing good service to demands and utilizing resources efficiently.

2.3.1 Resource Sharing Principles

There are two well-known resource sharing principles based on queuing theory [KLEI79]. The first principle is the law of large numbers, which simply states that a large number of demands will present a total load to a system of resources which is equal to the sum of the average requirements of each individual demand. Furthermore, this total load is a highly predictable quantity. Thus, even though an individual bursty source is relatively unpredictable in its moment-to-moment bandwidth requirements, if many of these bursty traffic sources are concentrated into a single queue, the bandwidth becomes more predictable. This should be compared with the fixed assignment case where the total resource capacity is equal to the sum of the individual peak demands rather than the sum of the average demands. Assigning resources according to average demands is clearly more desirable since more users can be accommodated. The smoothing effect of the law of large numbers is the key to the resolution of resource assignment in a bursty traffic environment.

The second principle is simply the scaling effect which states that if the resource capacity is fixed, then increasing that throughput by a factor will reduce the response time by the same factor. In other words, a single, high-speed server is often preferable to multiple servers whose aggregate service capacity is the same as that of the single server.

The combination of these two resource sharing principles indicates that it is preferable for large numbers of users to share large-capacity resources dynamically. In doing so, the combined gains from the smoothing effect of the law of large numbers as well as from the scaling effect are achieved. The key to realizing these gains is to design multiple access protocols that resolve channel access conflicts without excessive overhead.

2.3.2 The Global Queue

Since a single, high speed server is preferable to multiple servers whose aggregate service capacity is the same as that of the single server, the multiple access problem becomes that of forming a global queue of outstanding requests among a population of geographically distributed users. To form such a queue, there is a need to:

❑   identify the ready users who desire channel access;
❑   assign the channel to exactly one ready user according to some scheduling discipline.

The difficulty is that in wireless networks, the broadcast channel is also the only means for coordinating the distributed users since the information necessary for maintaining the global queue can only be exchanged using the channel itself.

2.3.3 Measuring Resource Usage

As explained in Section 2.1, the effective bandwidth requirements of a source depends critically on its statistical properties such as peak rate (source dependent) and QoS requirements (user dependent). However, the usage of a network resource may not be accurately assessed by a simple count of the number of bits transported by the network. For example, to provide acceptable performance to bursty sources with tight delay and loss requirements, it may be necessary to keep the average utilization of a link below 10%. On the other hand, for constant rate sources or sources able to tolerate substantial delays, it may be possible to push the average utilization well above 90%. Hence, there is a need to establish a measure of resource usage which adequately represents the trade-off between sources of different types, and accommodates changing characteristics and requirements.

There are two approaches to this issue. One approach insists that a user provide the network with a full statistical characterization of its traffic source, which is then policed by the network. Another approach recognizes the difficulty for a user to provide any information on traffic characteristics, but expects the network to cope nevertheless. Note that both approaches still derive the benefits of statistical multiplexing. They merely differ in how much characterization effort is required. A correct balance will necessarily involve tradeoffs between the user's uncertainty about traffic characteristics and the network's ability to statistically multiplex connections in an efficient manner.

2.3.4 Resource Sharing in Wideband Wireless Networks

When multiple users share a wireless channel, at any given time, a small fraction of users may experience high ISI due to multipath delay spread, while most will not. Resource sharing can be exploited to partially offset the effects of multipath, thereby increasing the coverage area with minimum decrease in throughput. In addition, the need to resort to multipath mitigation techniques (e.g., antenna diversity) is minimized.

With resource sharing, users transmit at some high rate under normal propagating conditions. When conditions deteriorate, the rate is lowered such that the BER objective is maintained. Although it takes longer to complete transmission at a lower rate, the number of users simultaneously slowing down is usually a small fraction of the total population. Thus, the overall throughput remains high.

2.3.5 Resource Reservation and Application Adaptation

Reservation and application adaptation are two approaches in maintaining orderly sharing of network resources among a variety of packet streams.

As the name implies, the reservation approach requires the resources to be allocated before any traffic is forwarded to the network, thereby insulating a packet stream from the adverse effects of sharing from other streams. The application actually knows the characteristics of its traffic and some specification of its required QoS. While the latter may not be difficult (e.g., an interactive voice application requires less than 100 ms end-to-end delay), characterizing the traffic of a multimedia application can be hard.

In application adaptation, the application first observes whether the network is able to handle its offered traffic as well as its desired QoS, and then adjusts its traffic characteristics accordingly. Applications that can adapt their QoS requirements and traffic characteristics according to available network resources in this manner can afford to be less accurate when making resource reservations.

2.3.6 Admission Control

In most procedures of connection admission, a contract is agreed between the user and the network. The contract specifies the statistical characteristics of the connection and the policing mechanisms that will enforce the contract.

In general, the network decides whether incoming reservation requests can be admitted based on two criteria:

❑   whether the requested QoS can be met given the specified traffic characteristics and the currently available network resources;
❑   whether admitting a request will affect the QoS promised to other traffic streams.

This requires the network to maintain state information.

Once admitted, traffic shaping and policing ensure that the QoS of the admitted stream is maintained. Traffic shaping is performed by the traffic source to enforce the traffic characteristics that has been specified to the network. Traffic policing is a performed by the network to ensure that a particular traffic source conforms to its stated traffic characteristics. Packets violating the agreed characteristics are either dropped or tagged for future discarding if they cause additional network congestion.

The admission decision can be based on a simple model of an unbuffered source [HUI88]. In this case, the probability of resource overload can be held below a desired level by requiring that the number of connections (n_j) accepted from sources of class j satisfies

$$\sum_{j=1}^{\infty} \alpha_j n_j \leq C$$

$$(2.2)$$

where C = capacity of the resource.

   α_j = effective bandwidth of each source of class j.

The effective bandwidth lies between the mean and the peak resource requirement of a source of class j. It depends on the characteristics of the source such as its burstiness and on the degree of statistical multiplexing possible at the source.

2.4 PERFORMANCE ANALYSIS

A consequence of the conservation law in queuing theory (see Appendix) is that the average delay performance of a multiple access protocol is independent of the order of service but depends mainly upon the time wasted for assigning channel access [KLEI76]. Thus, when these protocols are compared solely on average message delay performance for a given throughput level, the key problem is just the following: Whenever a channel is free and there are one or more ready users, how quickly can channel access be assigned to the user? The amount of time needed will be referred to as the channel assignment delay.

The general behavior of the average assignment delay versus throughput characteristic of a multiple access protocol is shown in Figure 2.1. Suppose T represents the average packet transmission time. The average channel assignment delay (d_1) under light traffic conditions determines the minimum delay (D_{min}) as follows:

$$D_{min} = T + d_1$$

(2.3)

On the other hand, the average channel assignment delay (d_2) under heavy traffic conditions determines the maximum throughput (S_{max}) as follows:

$$S_{max} = \frac{T}{T + d_2}$$

(2.4)

Both formulas may be adjusted slightly to account for channel propagation and other processing delays. To obtain the complete throughput-delay performance characteristic of a multiple access protocol requires a detailed mathematical analysis using queuing or Markov models. Generally, delay is more fairly distributed if users are served in the order of arrival into queue.

2.5 PERFORMANCE EVALUATION

Multiple access schemes for broadband wireless networks are evaluated according to various criteria. Desirable performance characteristics include:

- ❑ high bandwidth utilization and low packet delays;
- ❑ ability to support different traffic types, priorities, variable packet lengths, and variable delay constraints;
- ❑ robustness in terms of insensitivity to errors due to noise or other degradations;
- ❑ flexibility with respect to network growth.

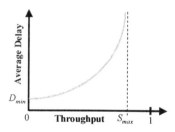

Figure 2.1: Throughput-delay performance of a multiple access protocol

It is important to realize that the performance measures of average message delay and average throughput are appropriate for a homogeneous population of users. Scheduling issues of fairness, discrimination, and priority traffic handling are also important considerations for broadband networks handling diverse traffic types with different performance goals (e.g., high throughput, low delay).

Four basic performance measures can be used to evaluate multiple access performance. These are efficiency, throughput, response time, and fairness.

2.5.1 Efficiency

Multiple access efficiency is defined as the maximum bandwidth that can carry end-user traffic while still ensuring stable operation. When multiple access protocols are employed in wireless networks, efficiency becomes a key consideration and is probably the most important measure from the network provider's point of view since it impacts network costs.

2.5.2 Throughput

While efficiency describes how well bandwidth resources are being managed, throughput characterizes how much bandwidth a particular user is able to receive from the network. The maximum end-user throughput under light traffic can be derived directly from the network efficiency when a single user transmits. Under heavy traffic, the average throughput is given in terms of the network efficiency and the number of transmitting users.

2.5.3 Response Time

Response time is the time for a newly generated message to travel from source to destination. It is probably the most important parameter measure from the end-user's point of view. Networks can successfully carry real-time traffic if the amount of jitter introduced by the network is small. Therefore, a small response time variance is just as important as a small average response time.

2.5.4 Fairness

Fairness refers to the use of channel resources by competing users. For wireless networks where limited bandwidth resources are involved, utilization is expected to be high and fairness is certainly a major issue. The network must be able to provide consistent QoS to all users when desired. This means that bandwidth allocation must be apportioned fairly and heavily-loaded users must prevented from monopolizing channel transmission time. In this case, the measure of fairness relates to performance expectations set by the application. Therefore, the comparison of network response time perceived by different end-users constitutes a suitable fairness measure.

2.6 IMPLEMENTATION CONSIDERATIONS

Depending on system requirements and the nature of the applications supported by the system, it is important to implement the appropriate multiple access technique. Several options exist, and their pros and cons are discussed in this section.

2.6.1 Centralized versus Distributed

The design of a wireless network can be simplified considerably if the information about the network and the intelligence to manage it are all collected in one location. A centralized base station can arbitrate access among contending users, provide a convenient access point to the backbone network, assign addresses and priority levels, monitor network load, manage forwarding of messages, and keep track of the current network topology. Centralized control also prevents wastage of channel resource in the event when user terminals malfunction or move out of a radio cell after making successful reservations for bandwidth.

Some distributed reservation protocols inevitably rely on communicating users being able to synchronize their transmissions according to their positions on a global queue. This requires users to exchange control (feedback) information, either explicitly or implicitly. Using this information, all users then execute independently, the same algorithm, resulting in some coordination in their actions. Clearly, it is essential that all users receive the same information regarding the demands placed on the system and its usage. The latter requirement is not always viable in an unpredictable wireless channel since a receiver may easily misinterpret the actual outcome of a transmission. Specifically, transmission errors and collisions may be hard to decipher. As a result, such protocols are liable to suffer from problems of deadlock arising from erroneous channel feedback.

Other benefits of a centralized configuration relate to indoor wireless operation. A strategically-located base station is able to minimize transmit power and deal with the problem of hidden users effectively. The physical range of transmission is usually greater than that provided by a distributed architecture.

2.6.2 Mobility versus Portability

A key requirement for wireless networks is the ability to handle both mobile and portable users. A portable user is one that is moved from location to location but is only used while in a fixed location. Mobile users actually access the network while in motion. User mobility requires radio cells to be properly overlapped so that continuous network access can be maintained. This implies that appropriate coordination between cells must be incorporated for seamless handoff.

The mobility management function (generally implemented at some level above the access layer), can provide useful location information to a multiple access scheme. By employing smart antennas, such location information can help to mitigate the effects of spatially distributed interference.

2.6.3 Integration with Higher Layer Functions

An efficient and reliable multiple access protocol for broadband wireless networks needs be closely integrated with higher-layer functions in addition to its integration with lower-layer functions. Such a protocol must be designed to support scalable or layered source coding approaches with layer-specific QoS requirements. Furthermore, in order to provide QoS guarantees, the multiple access technique must adapt to changes in interference conditions and user mobility. This can best be accomplished by integrating the multiple access function with information provided by lower and higher layers in the protocol stack.

SUMMARY

The principles of resource sharing and scheduling advocate the benefits of organizing message transmissions of individual queues from independent users into a single, cooperative, and statistical multiplexing global queue. Additional loss of resources is the cost for organizing separate demands. In wireless access protocol design, this cost will involve some combination of idle resources, collisions, and control overhead.

Metrics defined in terms of efficiency, throughput, response time, and fairness can be used to evaluate the performance of a multiple access protocol. In addition, the system implementation aspects of an access protocol can be evaluated based on the centralized or distributed nature of its architecture, the need to support mobility or portability, and its integration with information provided by lower and higher protocol layers.

BIBLIOGRAPHY

[ADAM93]   Adam, J., "Interactive Multimedia", *IEEE Spectrum*, March 1993, pp. 23 – 39.

[DUBO94]   Dubois, E., "Guest Editorial: Digital Video Communications", *IEEE Communications Magazine*, May 1994, p. 37.

[EPST95]   Epstein, B. and Schwartz, M., "Reservation Strategies for Multi-Media Traffic in a Wireless Environment", *Proceedings of the 45th IEEE Vehicular Technology Conference*, 1995, pp. 165 – 169.

[GOOD97] Goodman, D. and Raychaudhuri, D., *Mobile Multimedia Communications*, Plenum Press, 1997.

[HABI92] Habib, I. and Saadawi, T., "Multimedia Traffic Characteristics in Broadband Networks", *IEEE Communications Magazine*, July 1992, pp. 48 – 54.

[HONC97] Honcharenko, W., Kruys, J., Lee, D. and Shah, N., "Broadband Wireless Access", *IEEE Communications Magazine*, January 1997, pp. 20 – 26.

[HUI88] Hui, J., "Resource Allocation for Broadband Networks", *IEEE Journal on Selected Areas in Communications*, Vol. SAC- 6, No. 9, pp. 1598 – 1608.

[IEEE87] "Special Issue on the Performance Evaluation of Multiple Access Networks", *IEEE Journal on Selected Areas in Communications*, Vol. SAC-5, No. 6., July 1987.

[JABB94] Jabbari, B., *Worldwide Advances in Communications Network*, Plenum Press, 1994.

[JABB96] Jabbari, B., "Teletraffic Aspects of Evolving and Next Generation Wireless Communication Networks", *IEEE Personal Communications*, December 1996, pp. 4 – 10.

[JAYA95] Jayant, N., Ackland, B., Lawrence, V. and Rabiner, L., "Multimedia – From Vision to Reality", *AT&T Technical Journal*, September/October 1995, pp. 14 – 33.

[JORD93] Jordan,S. and Varaiya, P., "Throughput in Multiple Service, Multiple Resource Communication Networks", *IEEE Transactions on Communications*, Vol. 39, No. 8, August 1991, pp. 1216 – 1222.

[KARL96] Karlsson, G., "Asynchronous Transfer of Video", *IEEE Communications Magazine*, August 1996, pp. 118 – 126.

[KLEI76] Kleinrock, L., *Queueing Systems Volume 2: Computer Applications*, John Wiley, 1976.

[KLEI79] Kleinrock, L., "On Resource Sharing in a Distributed Communication Environment", *IEEE Communications Magazine*, January 1979, pp. 27 – 34.

[KLEI00] Kleinrock, L., "On Some Principles of Nomadic Computing and Multiaccess Communications", *IEEE Communications Magazine*, July 2000.

[KWOK92] Kwok, T., "Communications Requirements of Multimedia Applications: A Preliminary Study", *Proceedings of the IEEE International Conference on Selected Topics in Wireless Communications*, June 1992, pp. 138 – 142.

[LAM83]   Lam, S., "Multiple Access Protocols", appearing in [LAM84].

[LAM84]   Lam, S., *Principles of Communication and Networking Protocols*, IEEE Computer Society Press, 1984.

[LUIS98]   Luise, M. and Pupolin, S., *Broadband Wireless Communications: Transmission, Access, and Services*, Springer-Verlag, 1998.

[NAGH96]   Naghshineh, M. and Acampora, A., "QoS Provisioning in Micro-Cellular Networks Supporting Multiple Classes of Traffic", *ACM/Baltzer Journal on Wireless Networks*, October 1996, pp. 195 – 203.

[PANC93]   Pancha, P. and Zarki, M., "Bandwidth Requirements of Variable Bit Rate MPEG Sources in ATM Networks", *Proceedings of the IEEE INFOCOM*, 1993, pp. 902 – 909.

[PANC94]   Pancha, P. and Zarki, M., "MPEG Coding for Variable Bit Rate Video Transmission", *IEEE Communications Magazine*, May 1994, p. 54 – 66.

[RODR90]   Rodrigues, M., "Evaluating Performance of High Speed Multiple Access Protocols", *IEEE Network Magazine*, May 1990, appearing in [STAL94].

[SCHW95]   Schwartz, M., "Network Management and Control Issues in Multimedia Wireless Networks", *IEEE Personal Communications Magazine*, June 1995, pp. 8 – 16.

[SHEN92]   Sheng, S., Chandrakasan, A. and Brodersen, R., "A Portable Multimedia Terminal", *IEEE Communications Magazine*, Vol. 30, No. 12, December 1992, pp.64 – 74.

[SHRO94]   Shroff, N. and Schwartz, M., "Modeling VBR Video Over Networks End-to-End Using Deterministic Smoothing at the Source", *International Journal of Communications Systems*, Vol. 7, pp. 337 – 348.

[STAL94]   Stallings, W., *Advances in Local and Metropolitan Area Networks*, IEEE Computer Society Press, 1994.

[TEGE95]   Teger, S., "Multimedia – From Vision to Reality", *AT&T Technical Journal*, September/October 1995, pp. 4 – 13.

[TOBA93]   Tobagi, F., "Multimedia: The Challenge Behind the Vision", *Data Communications*, January 1993, pp. 61 – 67.

[WANG96]   Wang, Y. (editor), *Multimedia Communications and Video Coding*, Plenum Press, 1996.

Chapter 3

MULTIPLE ACCESS COMMUNICATIONS

Defining a multiple access protocol essentially implies the specification of a set of rules to be followed by each member of the user population in order to share a common bandwidth resource in a cooperative manner. The overall communications channel can be divided into subchannels which are then assigned to contending users. Typically, there are more users than the available subchannels but only a fraction of all users have packets to transmit at any given time. The central problem therefore, is to locate the users with data to send in order for these users to share the channel efficiently. However, the broadcast nature of the wireless channel poses a difficult issue for multiple access communications in that the success of a transmission is no longer independent of other transmissions. To make a transmission successful, interference must be avoided or at least controlled. Multiple simultaneous transmissions lead to collisions and corrupted signals. Thus, a multiple access protocol must resolve these access contentions among users and transform a multiple access network into a logical point-to-point network. The domains that contention resolution can be achieved include time, frequency, code, space, or some combination.

3.1 CHARACTERIZING THE ACCESS PROBLEM

Multiple access protocols are most beneficial when point-to-point and channelized communications become inefficient. An example is the case when there are a large number of users transmitting bursty data in an uncoordinated manner. The need for multiple access protocols also arises whenever a resource is to be shared and accessed by geographically distributed users and also, when there is a need for communication among independent users that form an interconnected network. For wireless networks, sharing of spectrum (channel bandwidth) is essential because radio spectrum is not only expensive but also inherently limited. The channel is a shared resource whose allocation is crucial for the efficient operation of any wireless network. This sharing task is made harder when disparate traffic types created by multimedia applications are to be transported across the network.

3.1.1 General Characteristics

The general characteristics of a multiple access network [ABRA93] are:

❑ The network contains independent users attempting to communicate using a single common channel or perhaps even multiple channels but the number of channels is much smaller than the number of users with data to send (the ready users). Note that a ready user is not a transmitting user. A ready user needs to check whether the channel is idle or not before it initiates transmission.

❑ At any given time, the number of ready users on the network is unknown and may change dynamically with time.

3.1.2 Classification

Besides their varying degrees of appropriateness to different communication environments, multiple access strategies differ by the static or dynamic nature of the bandwidth allocation procedure, the centralized or distributed nature of the decision-making process, the level of co-ordination among users, and the degree of adaptability of the algorithm to changing needs. Figure 3.1 shows how multiple access protocols can be classified according to the manner in which bandwidth is allocated. Many of these techniques were originally designed for deployment in terrestrial wireless and satellite packet switching networks.

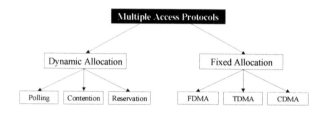

Figure 3.1: Classification of multiple access protocols

The main advantage of fixed assignment schemes is that each user is guaranteed a share of the link's bandwidth and as such, transmissions generally do not interfere with one another. The disadvantage is that channel resource is dedicated to users and unused bandwidth cannot be transferred from one user to another. Thus, while fixed assignment protocols have excellent throughput performance under steady, heavy traffic conditions, the delay performance when the traffic load is light and bursty can be quite poor. The obvious solution to this problem is to design a protocol that allocates channel resource only to users with data to transmit. A typical way to achieve this is to maintain a global queue of requests from users who wish to transmit. Unfortunately, such users must be identified and this information can only be exchanged through the channel itself. As a result, dynamic access schemes based on polling and reservation have been developed. Most of these schemes are very attractive since they offer reasonably low delay under light traffic and high throughput under heavy traffic. However, if the overhead involved in maintaining the global queue becomes large (e.g., due to long propagation delays), the performance of these schemes also degrades.

The class of contention protocols is useful when very few users wish to access the channel at any given time. In this case, rather than maintain a global queue, each ready user decides independently when to access the channel. These uncontrolled schemes are very easy to implement but pay the price in the form of wasted bandwidth due to collisions when one or more users access the channel at the same time. Such protocols usually have excellent delay performance under low traffic but the throughput degenerates quickly under high load. Contention protocols do not require the actual identities of the ready users to be known in advance. This feature is particularly important for wireless networks where mobile users can communicate from different locations.

Generally, the throughput-delay curves of fixed assignment and contention techniques intersect each other at a crossover point that determines which technique is better (Figure 3.2). This crossover point varies over a wider range of channel utilization and depends on specific network parameters. Clearly, a protocol that switches between contention and linear search at the crossover point achieves optimum performance.

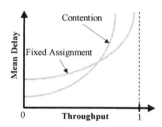

Figure 3.2: Fixed assignment versus contention

It is not the intention of this book to give an exhaustive survey on each and every devised method since much detailed information is available in the literature spanning over three decades. Useful surveys can be found in [ABRA93], [BERT92], [FRAN81], [KURO84], [LAM84], [ROM90], [TANE88] [TASA86], [TOBA89], and [TROP81].

3.3 FRAMEWORK FOR DISCUSSION

In order to provide a coherent framework for discussing the protocols in the later chapters, the characteristics of an ideal multiple access protocol and an ideal network are defined in Tables 3.1 and 3.2 respectively.

Table 3.1: Characteristics of an ideal multiple access protocol

Characteristic	Description
Immediate access	If channel is idle, immediate transmission is possible
Full channel utilization	Throughput of network is equal to the offered load up to an offered load of one
Predictable delay	Each transmission has a predictable delay which is minimum up to an offered load of one
Fair access	Each transmission request is met on a first-come-first-served basis
Distributed control	Each user manages its own requests and transmissions independently

Table 3.2: Characteristics of an ideal network

Characteristic	Description
Poisson arrivals	Packet arrivals are statistically independent of one another
No capture	A collision occurs when two or more packets overlap (partially and fully)
Equiprobable error	Errors due to channel noise affect all transmissions equally and are ignored
Immediate ternary feedback	Feedback at the end of each time slot indicating idle, success or collision
Collisions are resolved first	Collided packets are retransmitted before any newly generated packets
No buffering	New packets are blocked until the current packet is successfully transmitted

3.3.1 General Network Assumptions

Broadcast wireless networks can be classified under two main groups namely, fully-connected and multihop. In fully-connected networks, all users are in the same wireless coverage area and can therefore detect all activities of the channel. In multihop wireless networks, some users are unable to communicate directly with other users and must rely on one or more intermediate nodes to forward packets of information.

In some multiple access protocols, the users are time-synchronized so that the channel can be viewed as a sequence of discrete intervals (time slots). Each time slot can accommodate one data packet and packet transmission always start at the beginning of a slot. Minislots may also be interleaved with the data slots to accommodate short control packets.

3.3.2 Collisions

For contention protocols, it is possible to have overlapped transmissions from multiple users. Any overlap in packet transmission (no matter how minor) generally results in a collision and the destruction of all packets involved. However, if a single packet is transmitted, it is received with no errors. This assumption implies that collisions are the sole cause of packet errors and removes the noise and communication aspects from the multiple access problem, thereby allowing many contention protocols to be analyzed in the simplest context. Note that there is a significant difference between transmission errors and packet collisions. Transmission errors due to noise affect only a single user whereas collisions involve two or more users.

3.3.3 User Population

If multiple access protocols are modeled using traditional queuing theoretic models, problems arise since the service time (i.e., time to successfully transmit a packet) for a user is directly dependent on the queues of other users in the network [SILV82]. This results in a Markov chain with no simple closed form solution and can only be solved by numerical techniques. Such numerical techniques are only feasible for small networks.

It is common to assume an infinite population with no buffering capability so that closed form solutions can be obtained. This assumption considers each user to be able to successfully transmit its packet in a period of time that is small compared to the time it takes to generate a new packet. Thus, no queuing of packets ever takes place and all outstanding packets (i.e., those generated but not yet successfully transmitted) belong to different users. From a practical point of view, this is an unreasonable assumption. It is also not a reasonable assumption for fixed allocation multiple access protocols (e.g., FDMA, TDMA) since the capacity of such a system is zero when every user has a packet to transmit. Moreover, for such protocols, it is impossible to identify all users in an infinite population. Here, a multiple access protocol must search for packets directly, for example, examining the time of arrivals of each packet.

However, an infinite number of users is amenable to a finite user network with the additional possibility that a physical transmitter can sometimes send simultaneous, colliding packets. This implies that the assumption provides a worst case (upper) bound on a finite set of users and that the difference is only significant when two or more packets are waiting at the same transmitter [GALL85]. Since most contention protocols are unstable under high load, the analysis for such systems are most useful under conditions of light load where multiple packets rarely queue up at one transmitter. In this case, the performance of a finite set of transmitters is well approximated by the infinite set.

3.3.4 Propagation Delay

The propagation delay of a channel is always taken to be the maximum time for a signal to travel between any pair of transmitter and receiver in the network. Some protocols have a collision detection period, which is taken as approximately equal to the propagation delay. An important parameter in multiple access communications is the ratio of the propagation delay over the packet transmission time. This value of this ratio affects the performance of many access protocols.

3.3.5 Channel Feedback

The types of channel feedback play a crucial role in the design and development of time-slotted, contention access protocols. Typically, channel feedback allows users to learn immediately whether there has been a packet transmission at the end of the each predefined time slot. Such an immediate feedback model may also be generalized to the case where feedback is delayed although this complicates analysis with little benefit of insight.

The aim of many collision resolution protocols is to exploit feedback information to resolve or avoid conflicts due to simultaneous transmissions. Transmitters monitor the activity of the channel, obtain some kind of feedback based on past transmissions, and then schedule transmissions so as to maximize the achievable throughput and/or minimize the average delay for a given throughput.

There are three types of feedback. Users can learn whether there has been zero (idle), one (success) or more than one transmission (collision) has taken place in previous transmissions. There are two ways in which no packets can be successfully transmitted – no transmission or two or more transmissions. Thus, from an information theoretic point of view, up to one bit of information can be gained. However, there is only one way in which a packet can be successfully transmitted so no additional information is gained. Thus, a contention access protocol can learn more from failure than from success and it must risk failures to gain the required information [PIPP81]. In other words, a compromise is necessary between the desire to transmit a packet successfully in a given slot and the desire to gain information in order to transmit future packets successfully in later slots.

Due to channel errors, the receiver may misinterpret the actual outcome of a transmission. Interpretations of a conflict (collision) or a success slot as an idle slot belong to the class of erasures. Erasures can occur when mobile users are occasionally hidden from the receiver (e.g., due to physical obstacles or fading problems). Interpretations of an idle or a success slot as a conflict slot are classified under noise errors. Such errors are intrinsic in any communication channel. If capture effects are taken into account, then a conflict is interpreted as a success. Note that in this case, all other packets involved in the conflict, other than the captured packet are erased.

If all three types of feedback are available, then the channel is said to possess ternary feedback. However, it is also common to assume a binary channel. The various types of binary feedback include success/failure (informs users whether or not there was a successful transmission in the previous slot), collision/no collision (informs users whether or not there was a collision in the previous slot), and something/nothing (informs users whether or not the previous slot was empty).

The collision/no collision feedback is the most informative whereas the success/failure feedback provides the poorest kind of binary feedback. The main problem is attributed to the fact that when a failure feedback is received, the users who transmitted in the corresponding slot recognize it as a collision while those who did not transmit cannot distinguish between a collision and an idle channel (i.e., between collision and channel noise) [PATE89]. However, this type of feedback is common and arises when explicit channel sensing is not possible. For example, in spread spectrum systems, a collision of two or more transmitted signals results in a noise-like waveform that is hard to discriminate reliably from pure channel noise.

3.3.6 Full and Limited Sensing

Another important factor affecting the performance of multiple access protocols is the manner which users are permitted to sense the channel. In limited sensing, each user is required to observe the channel feedback, from the time is generated to the time it is successfully transmitted. On the other hand, full sensing requires the user to monitor the channel continuously, regardless of whether it has a ready packet to send or not. Contention access protocols work best when all transmitters are able to monitor the channel at all times.

SUMMARY

This chapter has discussed how a common broadcast medium can be shared among many contending users. Multiple access protocols differ primarily by the amount of coordination needed to control potentially conflicting packet transmissions. At one extreme is random access where no coordination is provided and packet collisions are possible. At the other end of the spectrum, the class of fixed assignment access protocols eliminates collisions entirely but pay the price of additional overhead required for scheduling user access. Hybrid access protocols between these two extremes exist. While these protocols attempt to combine the advantages of random and fixed access, they also suffer the combined drawbacks and overhead of both classes of access schemes. Among the many factors that determine the performance of an access protocol include the propagation delay/packet transmission time ratio, the message arrival process, the types of feedback information available, the user population, and the ability of the user to sense the activities in the network.

BIBLIOGRAPHY

[ABRA93]   Abramson, N., *Multiple Access Communications*, IEEE Press, 1993.

[BERT92]   Bertsekas, D. and Gallager, R., *Data Networks*, Prentice Hall, 1992.

[CHOU83]   Chou, W, *Computer Communications Volume 1: Principles*, Prentice Hall, 1983.

[CIDO87]   Cidon, I. and Sidi, M., "Erasures and Noise in Splitting Multiple Access Algorithms", *IEEE Transactions on Information Theory*, Vol. 33, No. 1, January 1987, pp. 132 – 140.

[CIDO88]   Cidon, I., Kodesh, H. and Sidi, M., "Erasure, Capture and Random Power Level Selection in Multiple-Access Systems", *IEEE Transactions on Communications*, Vol. 36, No. 3, March 1988, pp. 263 – 271.

[ERIK93]   Eriksson, H., Gudmundson, G., Skold, J., Ugland, J. and Willars, P., "Multiple Access Options for Cellular-Based Personal Communications", *Proceedings of IEEE PIMRC*, 1993, pp. 957 – 962.

[FRAN81]   Franta, W. and Chlamtac, I., *Local Networks*, Lexington Books, 1981.

[GALL85]   Gallager, R., "A Perspective on Multiaccess Channels", *IEEE Transactions on Information Theory*, Vol. IT - 31, No. 2, March 1985, pp. 124 – 142.

[HUMB86]   Humblet, P., "On the Throughput of Channel Access Algorithms with Limited Sensing", *IEEE Transactions on Communications*, Vol. COM-34, No. 4, April 1986, pp. 345 – 350.

[KLEI77]   Kleinrock, L., "Performance of Distributed Multiaccess Computer Communication Systems", *Proceedings of the IFIP Congress*, 1977, pp. 547 – 552.

[KLEI80]   Kleinrock, L. and Yemini Y., "Interfering Queuing Processes in Packet-Switched Broadcast Communication", *Information Processing 80*, IFIP, North-Holland Publishing Company, 1980, pp. 557 – 562.

[KLEI82]   Kleinrock, L., "A Decade of Network Development", *Journal of Telecommunication Networks*, 1982, appearing in [LAM84], pp. 50 – 60.

[KURO84]   Kurose, J., Schwartz, M. and Yemini, Y., "Multiple Access Protocols and Time Constrained Communication", *Computing Surveys*, March 1984, pp. 43 – 70.

[LAM83]   Lam, S, "Multiple Access Protocols", appearing in [CHOU83] and [LAM84].

[LAM84]   Lam, S., *Principles of Communication and Networking Protocols*, IEEE Computer Society Press, 1984.

[LI87]   Li, V., "Multiple Access Communication Networks", *IEEE Communications Magazine*, Vol. 25, No. 6, June 1987, pp. 41 – 48.

[MEHR84]   Mehravari, N. and Berger, T., "Poisson Multiple-Access Contention with Binary Feedback", *IEEE Transactions on Information Theory*, Vol. IT-30, No. 5, September 1984, pp. 745 – 751.

[MOLL82]   Molle, M., "On the Capacity of Infinite Population Multiple Access Protocols", *IEEE Transactions on Information Theory*, Vol. IT-28, No. 3, May 1992, pp. 396 – 401.

[PATE89]   Paterakis, M. and Papantoni-Kazakos, P., "A Limited Sensing Random-Access Algorithm with Binary Success-Failure Feedback", *IEEE Transactions on Communications*, Vol. COM-37, No. 5, May 1989, pp. 526 – 529.

[PEYR99]   Peyravi, H., "Medium Access Control Protocols Performance in Satellite Communications", *IEEE Communications Magazine*, Vol. 37, No. 3, March 1999, pp. 62 – 71.

[PIPP81]   Pippenger, N., "Bounds on the Performance of Protocols for a Multiple-Access Broadcast Channel", *IEEE Transactions on Information Theory*, Vol. IT-27, No. 2, March 1991, pp. 145 – 151.

[ROM90]   Rom, R. and Sidi, M., *Multiple Access Protocols: Performance and Analysis*, Springer-Verlag, 1990.

[SILV82]   Silvester, J. and Lee, I., "Performance Modeling of Multi-Access Computer Communication Networks", *Proceedings of IEEE GLOBECOM*, 1982, pp. 377 – 382.

[SKIR81]   Skwirzynski, J., ed., *New Concepts in Multi-User Communication*, Nato Advanced Study Institutes Series, Sijthoff and Noordhoff International Publishers, The Netherlands, 1981.

[SUNS89]   Sunshine, C. (editor), *Computer Network Architectures and Protocols*, Plenum Press, 1989.

[TANE88]   Tanenbaum, A., *Computer Networks*, Prentice Hall, 1988.

[TASA84]   Tasaka, S., "Multiple-Access Protocols for Satellite Packet Communication Networks: A Performance Comparison", *Proceedings of the IEEE*, November 1984, pp. 1573 – 1582.

[TASA86]   Tasaka, S., *Performance Analysis of Multiple Access Protocols*, MIT Press, 1986.

[TOBA89]   Tobagi, F., "Multiple Access Link Control", appearing in [SUNS89].

[TROP81]   Tropper, C., *Local Computer Network Technologies*, Academic Press, 1981.

Chapter 4

FIXED ALLOCATION ACCESS PROTOCOLS

Fixed allocation schemes are practical in a network where users generate traffic in a fairly regular and predictable fashion, and the set of transmitters is known and small. The channel can be divided by providing a portion of the bandwidth to every user all the time as done in Frequency Division Multiple Access or providing the entire bandwidth to a single user for a fraction of the time as done in Time Division Multiple Access. An important advantage of fixed allocation access protocols is the ability to control the delay of packet transmission, which can be essential for real-time applications. This advantage is offset by the difficulty in adding new users and deleting old users.

4.1 FREQUENCY DIVISION MULTIPLE ACCESS

FDMA requires the total bandwidth to be divided into a number of disjoint frequency subchannels. Each subchannel is assigned on demand to users who request service. When all channels are occupied, the system has reached its system capacity and no further users may be accommodated.

For reasons of spectral efficiency, the transmission rate on a single FDMA channel is usually close to the maximum rate required by the user. Hence, FDMA is suitable for users with nonbursty and predictable traffic. If users generate unequal amounts of traffic, FDMA can be modified to assign bandwidth in proportion to the traffic produced by each user. For example, increased data rates can be achieved by concatenating a number of fixed FDMA channels together. This requires either duplicated RF front-ends or some form of multitone decoding facility. An alternative is to divide the overall bandwidth into a range of channels with unequal bandwidth. The users, however, will need to have different receiver filter bandwidths. This can be achieved using programmable DSP filter implementations.

4.1.1 Disadvantages

To reduce interference, guard bands are necessary between adjacent FDMA subchannels. These guard bands cannot be used for information transfer and therefore, waste a fraction of the available bandwidth. FDMA is also characterized by a lack of flexibility in accommodating changes to the allocated bandwidth as well as a lack of broadcast capability. Furthermore, non-linear power amplification (using Class-C amplifiers) can cause intermodulation distortion when many users transmit simultaneously.

4.1.2 Implementation

FDMA requires the construction of accurate frequency sources and highly selective bandpass filters. Such filters are necessary for receivers to reject interference from adjacent frequency channels.

In FDMA/FDD systems, the user is assigned a pair of frequencies, one for the uplink channel and the other for the downlink channel. FDMA can also operate with TDD. In this case, only a single frequency band is required. A time-frame structure provides one half for transmission and the other half for reception.

4.2 TIME DIVISION MULTIPLE ACCESS

In TDMA, channel access is divided into disjoint time slots which are assigned alternately among a fixed set of users. Unused time slots result in wasted bandwidth. Unlike FDMA, each user transmits at rate equivalent to the full network bandwidth. However, transmission time is restricted to the duration of the time slot. When all time slots are occupied, no additional users can be accommodated. A group of time slots usually form a frame that is repeated periodically (Figure 4.1). Such reuse in time is often combined with frequency reuse in cellular TDMA systems.

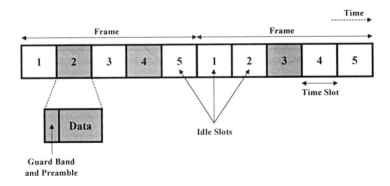

Figure 4.1: Operation of TDMA

4.2.1 Disadvantages

In TDMA, a fixed frame format must be defined and maintained. Users must observe a common time reference in order to determine the frame boundaries and the start of its own time slot within a frame. The high data rates imply that TDMA receivers must also be capable of rapid burst synchronization. If carrier and clock synchronization cannot be acquired quickly enough, then the preamble interval at the start of the transmission burst must be increased. Since acquisition and synchronization must be performed by the receiver for each transmission burst, the overhead for each burst can be substantial and this reduces the overall efficiency. For example, 30% of the overall data rate in GSM, PHS and DECT systems, and 20% of the data rate in IS-54 are absorbed by such overhead. Furthermore, the need to transmit bursts of information requires a high peak power requirement even though the average transmitter power of a TDMA user is much lower since the transmitter does not operate continuously and is idle for most of the frame duration.

Just as FDMA requires guard bands to avoid interference, TDMA systems must be equipped with sufficient guard times, which constitute another form of overhead. Guard times are used to compensate for [LEE94]:

- ❑ timing inaccuracies due to clock instability;
- ❑ delay spread;
- ❑ propagation delay;
- ❑ transient response of the pulsed signal.

For variable packet lengths, a compromise value of the time slot (which satisfies delay requirements while providing reasonable channel efficiency) must be chosen. This can be difficult if the variance in packet length is high. As an example, many multiple access schemes currently deployed in mobile cellular networks employ a fixed TDMA frame structure. Such schemes are optimized for continuous traffic rather than bursty traffic. To cater for multimedia traffic, the selection of a suitable frame structure is not an easy task since there is little knowledge of the traffic mix [HUI90]. If time slots are chosen to match the largest packet lengths, slot times that are under-utilized by short packets must be padded out to fill up the slots. On the other hand, if shorter slot lengths are used, more overhead per packet results [JACO78]. For some TDMA schemes, long frames are necessary in order to maximize the bandwidth utilization at high traffic loads but this is done at the expense of increasing the delay at low loads.

4.2.2 Implementation

The implementation of TDMA systems requires a compromise between the length of the time slot and the guard time. In micro- or pico-cellular networks, the guard time may be insignificant due to the short range. Generally, frequency, phase, bit timing, time slot, and frame synchronization in TDMA systems can be acquired using a preamble of roughly between 100 to 200 bit times for each burst.

In TDMA/TDD systems, half the time slots are used for the forward link and the other half for the reverse link. TDMA/FDD systems require two different frequency subbands in the forward and reverse directions although the same frame structure can be employed in both directions.

4.3 COMPARISON OF FDMA AND TDMA

The primary difference between TDMA and FDMA systems is that TDMA requires digital signals while FDMA can use both analog and digital signals. Another major distinguishing feature is that TDMA users transmit using the full bandwidth whereas FDMA users transmit using a fraction of the total bandwidth.

Due to the need for synchronization, TDMA is generally more complex to implement than FDMA. However, it has an important advantage of connectivity where all receivers listen to the same channel which senders transmit at different times. During this listening period, the user may use small-scale antenna diversity to choose the antenna with the best signal quality for the up-coming time slot. In addition, a common channel allows convenient information broadcast.

Through statistical multiplexing, TDMA systems utilize the available bandwidth efficiently. Since only one user transmits at any one time, TDMA is not prone to intermodulation distortion. The user data rate in TDMA can be made variable simply by allocating a different number of time slots on the same carrier without any hardware changes. This time division approach of separating user transmission is clearly more effective than the use of bandpass filters.

4.4 PERFORMANCE EVALUATION

In terms of delay performance, TDMA is superior to FDMA since the packet delay in FDMA is typically larger than TDMA. This applies to both multipacket messages and nonpreemptive priority queues [BERT92]. It is instructive to develop some analytical models that provide insights into the fundamental characteristics of FDMA and TDMA systems. These models should be compared with the statistical multiplexing case, which provides a lower bound to the delay performance.

Suppose N users, each with infinite buffer length, transmit packets of constant length (L bits). The packet arriving rate is λ packets/s. The packets are transmitted at an output data rate of either R bits/s or NR bits. For example, in TDMA, packets are transmitted using the full channel bandwidth of NR bits/s whereas FDMA systems transmit packets at a data rate of R bits/s.

4.4.1 Frequency Division Multiple Access

The queuing model for FDMA is shown in Figure 4.2. Using equation A.11 in the Appendix, the waiting time in the queue (W_{FDMA}) and the overall delay (D_{FDMA}) can be computed as follows.

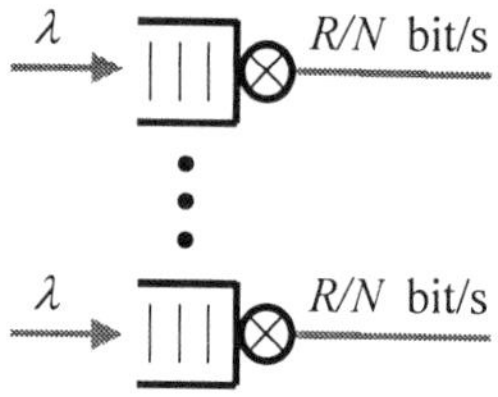

Figure 4.2: FDMA model

$$W_{FDMA} = \frac{\lambda\left(\dfrac{L}{R/N}\right)^2}{2(1-\rho)} = \frac{\lambda N^2 L^2}{2R^2(1-\rho)}$$

(4.1)

$$D_{FDMA} = \frac{L}{R/N} + W_{FDMA} = \frac{NL}{R} + W_{FDMA}$$

(4.2)

For the queue to be stable with finite delay, the input packet arrival rate must not exceed the output transmission rate. Thus,

$$\rho = \lambda \times \frac{L}{R/N} = \frac{N\lambda L}{R} < 1$$

(4.3)

4.4.2 Time Division Multiple Access

As illustrated in Figure 4.3, the delay in TDMA comprises two components:

☐ waiting time for the start of the time slot;
☐ waiting time for packets that arrived earlier in the same queue to be transmitted.

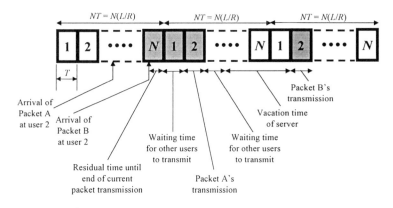

Figure 4.3: TDMA model

Using equation A.14, the waiting time in the queue (W_{TDMA}) is given by:

$$W_{TDMA} = \frac{\lambda\left(\frac{L}{R/N}\right)^2}{2(1-\rho)} + \frac{\left(\frac{L}{R/N}\right)^2}{2\left(\frac{L}{R/N}\right)} = \frac{\lambda N^2 L^2}{2R^2(1-\rho)} + \frac{NL}{2R}$$

(4.4)

The second term in equation 4.4 represents the fact that on the average, an arriving packet has to wait half the frame length (NL/R) before it is transmitted in its own time slot.

$$\therefore W_{TDMA} = W_{FDMA} + \frac{NL}{2R} > W_{FDMA}$$

(4.5)

The overall delay (D_{TDMA}) becomes:

$$D_{TDMA} = \frac{L}{R} + W_{TDMA} = \frac{L}{R} + \frac{NL}{2R} + D_{FDMA} - \frac{NL}{R} = \frac{L}{R} - \frac{NL}{2R} + D_{FDMA}$$

(4.6)

$$\therefore D_{FDMA} > D_{TDMA} \Rightarrow \frac{NL}{2R} - \frac{L}{R} > 0 \Rightarrow N > 2$$

(4.7)

Thus, for three or more users, the overall delay in FDMA is greater than TDMA (equation 4.7). The waiting time in TDMA, however, is greater than FDMA regardless of the number of users (equation 4.5). For $N = 1$, FDMA gives a lower overall delay since the half frame wait before transmission in TDMA is not needed in FDMA.

For stable queues,

$$\rho = N\lambda \times \frac{L}{R} = \frac{N\lambda L}{R} < 1$$

(4.8)

4.4.3 Statistical Multiplexing

The queuing model for statistical multiplexing (STDM) is depicted in Figure 4.4.

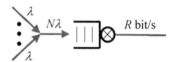

Figure 4.4: STDM model

The waiting time in the queue (W_{STDM}) and the overall delay (D_{STDM}) can be computed as follows:

$$W_{STDM} = \frac{N\lambda \left(\frac{L}{R}\right)^2}{2(1-\rho)} = \frac{N\lambda L^2}{2R^2(1-\rho)} = \frac{1}{N}W_{FDMA} < W_{FDMA} < W_{TDMA}$$

(4.9)

$$D_{STDM} = \frac{L}{R} + W_{STDM} = \frac{L}{R} + \frac{1}{N}\left(D_{FDMA} - \frac{NL}{R}\right)$$

(4.10)

$$\therefore D_{STDM} = \frac{D_{FDMA}}{N} < D_{FDMA}$$

(4.11)

Substituting equation 4.6,

$$D_{STDM} = \frac{1}{N}\left(D_{TDMA} - \frac{L}{R} + \frac{NL}{2R} \right)$$

(4.12)

$$D_{TDMA} - D_{STDM} = \frac{1}{N}\left[(N-1)D_{TDMA} + \frac{L}{R} - \frac{NL}{2R} \right]$$

$$= \frac{1}{N}\left[(N-1)\left(\frac{\lambda N^2 L^2}{2R^2(1-\rho)} + \frac{NL}{2R} + \frac{L}{R} \right) + \frac{L}{R} - \frac{NL}{2R} \right] = \frac{N\lambda(N-1)L^2}{2R^2(1-\rho)} + \frac{NL}{2R} > 0$$

(4.13)

$$\therefore D_{STDM} < D_{TDMA}$$

(4.14)

For stable queues,

$$\rho = N\lambda \times \frac{L}{NR} = \frac{\lambda L}{R} < 1$$

(4.15)

Note that unlike statistical multiplexing, the delay performance of TDMA and FDMA increases monotonically with N, even when the total input rate is very small.

4.5 POLLING PROTOCOLS

Polling protocols are normally useful for networks with short propagation delay (e.g., indoor wireless networks). A key characteristic of polling protocols is that they impose rigid registration procedures.

In conventional polling, the identities of all users in the network are known to a central controller which sends polling messages to the users, one at a time. If a polled user has a packet to transmit, it proceeds; otherwise, it sends back a short negative reply (or go-ahead message) to the controller. The controller then polls the next user in sequence. Polling requires this cyclic exchange of control messages between the controller and the users.

Each polling message contains the address of a single user. All users listen to the polling message and only the user that identifies its own address responds. The number of polling messages needed to find out the status of all users (ready or idle) is linearly dependent on the total number of users. This overhead is independent of the actual number of ready users present. Thus, although polling provides high reliability, it is difficult to achieve high channel efficiency unless:

❑   the round-trip propagation delay is short;
❑   the overhead due to polling messages is low;
❑   the total number of users is small.

An important advantage of polling is that it integrates well with the Logical Link Control (LLC) sublayer associated with LANs and is compatible with popular data link control protocols like ISO's High-level Data Link Control (HDLC). In addition to powerful error control procedures, HDLC provides an effective, window-based mechanism for performing flow (congestion) control. This can be critical for systems handling multimedia traffic with wide-ranging data rates and stringent delay constraints.

Other explicit forms of polling such as hub-polling or token-passing exist but the reliability of these methods are poor in a wireless environment since the loss of the token due to outages in received signal strength is frequent under fading wireless channels. Polling schemes when used in conjunction with an adaptive array of spatially diverse antenna elements at each base station of a wireless network provides excellent mitigation against multipath fading and co-channel interference [ACAM98].

4.5.1 Performance Analysis

A parameter unique to polling schemes is the walk time which represents the time required to transfer a polling message from one user to another. It puts an irreducible minimum time on the access delay experienced by users. Suppose J is the total walk time for N users and T_i is the average packet transmission time for user i. If the arrival rate for user i is λ_i, then the average polling cycle time is:

$$T_c = J + \sum_{i=1}^{N} T_i = J + \rho T_c$$

(4.16)

where

$$\rho = \sum_{i=1}^{N} \lambda_i T_i$$

(4.17)

$$\therefore T_c = \frac{J}{1-\rho}$$

(4.18)

The average access delay is given as [SCHW87]:

$$W_{POLLING} = \frac{T_c}{2}\left(1 - \frac{\rho}{N}\right) + \frac{N\lambda\overline{T^2}}{2(1-\rho)}$$

(4.19)

4.5.2 Probing

With conventional polling, when the network is lightly loaded, it is still necessary to poll every user even though most users are idle. Hence, the average delay is determined by the polling overhead and not by the number of ready users. The amount of polling overhead is critically dependent on the time to transfer a polling message from one user to another.

For a lightly loaded network, the method of probing can be used. The basic idea is to poll a group of users all at once and if none responds, then the whole group is eliminated. However, when one or more ready user responds, collisions occur. The controller then breaks down the user population into smaller subsets and repeats the same probing procedure (Figure 4.5). This process continues until a single ready user is isolated. The address of a group is chosen to be the common prefix of the user addresses in that group. For example, for a group of two users whose addresses are

010 and 011, the group address is 01. If a group comprises four users whose addresses are 100, 101, 110, and 111, then the group address is 1. Note that for an n-bit address, the maximum collision resolution interval is $2^{n+1} - 1$ and this occurs when all users wish to transmit. The collision resolution interval is therefore variable. It depends on the number of users involved in a collision as well as on their addresses.

4.5.3 Adaptive Probing

Suppose the number of users in the network is 2^n. Then each user is assigned an n-bit address. Let a cycle be defined as the time required for polling and transmission of all packets generated in the preceding cycle. If a single user has a packet to transmit, probing requires one query per cycle as opposed to 2^n for conventional polling. However, if all users have packets, probing requires $2^{n+1} - 1$ queries as opposed to 2^n for conventional polling.

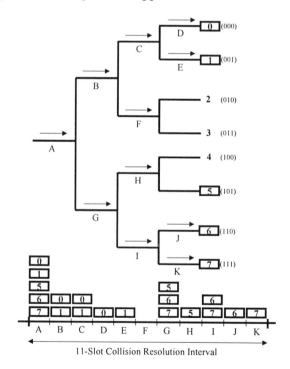

Figure 4.5: Probing

Clearly, probing becomes inefficient when the number of ready users is large. For improved performance in a heavily loaded network, the probing technique can be made adaptive. The controller starts a cycle by probing smaller groups as the probability of users having packets increases. In particular, the size of these groups may be considered as a function of the duration of the preceding polling cycle. Simulation of the adaptive probing strategy has shown that this scheme is always superior to conventional polling in that its average cycle time is always shorter [HAYE78], [HAYE84].

4.6 SERVICE DISCIPLINES IN POLLING SYSTEMS

A unique feature of polling systems is the ability to ensure fairness among ready users through prioritized polling and strict packet delay control. For example, heavily-loaded users can be prevented from monopolizing channel transmission time. Parameters such as polling order and service duration can be used to prioritize users and improve system performance.

The order by which the server visits the users can be determined either prior to operation (static order) or during operation (dynamic order) [LEVY90]. The advantage of dynamic orders is that they can adapt to the actual system state and thus, can be used to optimize its performance.

Limited policies are effective in controlling the service duration. They assign proper limits on the maximum number of packets that each user can transmit during one visit of the server. Such policies can be deterministic, probabilistic or fractional.

SUMMARY

In fixed assignment access protocols, the minimum delay increases in direct proportion to the number of users, regardless of whether these users are transmitting or not. Thus, more users can share a given channel bandwidth in such schemes provided all users are willing to tolerate greater delay. A key characteristic of polling protocols is the reliability in providing channel access. Such protocols are attractive for short-range wireless networks with a small number of users that generate stable and predictable traffic patterns.

BIBLIOGRAPHY

[ACAM98] Acampora, A., Krishnamurthy, S. and Zorzi, M., "Media Access Protocols for use with Smart Array Antennas to Enable Wireless Multimedia Applications", appearing in [LUIS98].

[BERR92] Berry, L. and Bose, S., "Traffic Capacity and Access Control", appearing in [STEE92], pp. 145 – 188.

[BERT92] Bertsekas, D. and Gallager, R., *Data Networks*, Prentice Hall, 1992.

[FALC95] Falconer, D., Adachi, F. and Gudmundson, B., "Time Division Multiple Access Methods for Wireless Personal Communications", *IEEE Communications Magazine*, January 1995, pp. 50 – 57.

[HAYE78] Hayes, J., "An Adaptive Technique for Local Distribution", *IEEE Transactions on Communications*, Vol. COM-26, August 1978, pp. 1178 – 1186.

[HAYE81] Hayes, J., "Local Distribution in Computer Communications", *IEEE Communications Magazine*, March 1981, pp. 6 – 14.

[HAYE84] Hayes, J., *Modeling and Analysis of Computer Communications Networks*, Plenum Press, 1984.

[HUI90]   Hui, J., *Switching and Traffic Theory for Integrated Broadband Networks*, Kluwer Publishers, 1990.

[JACO78] Jacobs, I., Binder, R. and Hoversten, E., "General Purpose Packet Satellite Networks", *Proceedings of the IEEE*, Vol. 66, No. 11, November 1978, pp. 1448 – 1467.

[KLEI90] Kleinrock, L. and Levy, H., "The Analysis of Random Polling Systems", *Operations Research*, Vol. 36, No. 5, September-October 1988, pp. 716 – 732.

[LAM77]   Lam, S., "Delay Analysis of a Time Division Multiple Access (TDMA) Channel", *IEEE Transactions on Communications*, Vol. COM-25, No. 12., December 1977, pp. 1489 – 1494.

[LAMA93] LaMaire, R, Krishna, A and Ahmadi, H, "Analysis of a Wireless MAC Protocol with Client-Server Traffic", *Proceedings of the IEEE INFOCOM*, 1993, pp. 429 – 436.

[LAMA94] LaMaire, R., Krishna, A. and Ahmadi, H., "Analysis of a Wireless MAC Protocol with Client-Server Traffic and Capture", *IEEE Journal on Selected Areas in Communications*, Vol. 12, No. 8, October 1994, pp. 1299 – 1237.

[LEE94]   Lee, E. and Messerschmitt, D., *Digital Communications*, Kluwer Academic Publishers, 1994.

[LEVY90] Levy, H. and Sidi, M., "Polling Systems: Applications, Modeling and Optimization", *IEEE Transactions on Communications*, Vol. 38, No. 10, October 1990, pp 1750 – 1760.

[LUIS97] Luise, M. and Pupolin, S., *Broadband Wireless Communications: Transmission, Access, and Services*, Springer-Verlag, 1998.

[MEHR84] Mehravari, N., "TDMA in a Random-Access Environment: An Overview", *IEEE Communications Magazine*, Vol. 22, No. 11, pp. 54 – 59.

[RUBI79] Rubin, I., "Message Delays in FDMA and TDMA Communication Channels", *IEEE Transactions on Communications*, Vol. COM-27, No. 5, May 1979, pp. 769 – 777.

[SALE91] Saleh, A., Rustako, A., Cimini, L., Owens, G. and Roman, R., "An Experimental TDMA Indoor Radio Communications System Using Slow Frequency Hopping and Coding", *IEEE Transactions on Communications*, Vol. 39, No. 1, January 1991, pp. 152 – 162.

[SCHW87] Schwartz, M., *Telecommunication Networks*, Addison Wesley, 1987.

[SPRA77] Spragins, J., "Simple Derivation of Queueing Formulas for Loop Systems", *IEEE Transactions on Communications*, Vol. COM-25, No. 5, April 1977, pp. 446 – 448.

[SPRA91] Spragins, J., *Telecommunications: Protocols and Design*, Addison Wesley, 1991.

[STEE92] Steele, R., *Mobile Radio Communications*, Pentech Press, 1992.

[STUC85] Stuck, B. and Arthurs, E., *A Computer Communications Network Performance Analysis Primer*, Prentice Hall, 1985.

[TAKA86] Takagi, H., *Analysis of Polling Systems*, MIT Press, 1986.

[TAKA88] Takagi, H., "Queueing Analysis for Polling Protocols", *Computing Surveys*, March 1988.

[TAKA90] Takagi, H. (editor), *Stochastic Analysis of Computer and Communication Systems*, Elsevier Science Publishers, 1990.

[ZHAN91] Zhang, Z. and Acampora, A., "Performance of a Modified Polling Strategy for Broadband Wireless LANs in a Harsh Fading Environment", *Proceedings of the IEEE GLOBECOM*, 1991, pp. 1141 – 1146.

Chapter 5

CONTENTION PROTOCOLS

Contention schemes provide access to bandwidth resources without imposing predictable or scheduled time for any user to transmit. The virtual elimination of network coordination makes such schemes extremely easy to implement since users can join or depart from the network with ease. This advantage is the main reason why wired Ethernet networks employing contention schemes are far more dominant than contention-free token ring networks. Such a feature is also of paramount importance in a wireless network where mobile users roam about freely.

5.1 GENERAL CHARACTERISTICS

Contention techniques allow quick access to the network when the traffic load is light. They typically cater for broadcast networks with a large or varying number of users with bursty traffic requirements. This means that most of the users have nothing to send most of the time and only a few are ready. For traffic that require synchronous (periodic) receiver playback (e.g., voice and video), additional buffers must be employed to compensate for the delay variability in contention protocols.

Unlike polling disciplines, the overhead incurred by contention protocols for assigning channel access to ready users is dependent only on the number of ready users and not the total number of users in the network. As such, the response time of contention protocols can be far superior to polling protocols in a lightly loaded network.

Collisions and subsequent retransmissions are the price of giving many users uncoordinated, random access to the same channel. Collisions:

❑   limit the amount of information (throughput) that can be transmitted through the channel;
❑   give rise to a random order of service;
❑   introduce variable packet delays (delay jitter).

A far more significant disadvantage of contention protocols is that statistical fluctuations in the traffic load may cause the channel to drift into an unstable, saturation state where the channel is filled with collisions and zero throughput results [CARL75], [LAM75]. Hence, additional control algorithms are needed to prevent channel saturation and these mechanisms attempt to improve throughput by reducing either the number of collisions or the length of the collision intervals. They are usually based on one or more of the following:

❑   reducing the probability of retransmitting a collided message;
❑   revoking the access rights of some users for a period of time.

5.2 ALOHA PROTOCOLS

Unslotted (pure) ALOHA and slotted ALOHA are contention protocols that are widely employed in wireless networks. Spread spectrum-based systems such as spread-ALOHA and disciplined-ALOHA have also been developed. The effects of excess capacity, variable slot length, and power capture on the ALOHA system have been studied extensively. In this section, several ALOHA-based protocols are discussed.

5.2.1 Unslotted ALOHA

In unslotted ALOHA, a user transmits a new packet at the instant it is generated, disregarding the activities of other users in the network. Because users are not synchronized or coordinated in any way, such a protocol is extremely simple to implement. Each successfully received packet is individually acknowledged, either through an explicit acknowledgment packet sent by a central controller/receiver or simply by allowing the user to listen to the channel for feedback information.

When two or more packets overlap in time (partially or fully), a collision occurs. In most situations, all collided packets are destroyed and must be retransmitted. Clearly, the duration of a collision interval in unslotted ALOHA is variable and can extend to several packet lengths (Figure 5.1). For packets of fixed-length L, the vulnerable period corresponding to the probability of a collision is $2L$ (Figure 5.2). If a propagation delay of τ is included, then the vulnerable period becomes $2(L + \tau)$.

5.2.2 Slotted ALOHA

The efficiency of unslotted ALOHA can be improved by imposing a little more cooperation among ready users. The slotted ALOHA restricts each packet to be of fixed-length that fits into a time slot. Packet transmissions are synchronized to arrive at the intended receivers at specific slot boundaries [ROBE75]. Packets arriving after a slot boundary have to be delayed until the next slot before transmission. In this way, collisions due to partial overlaps are avoided and the maximum collision period is reduced to just the duration of a single time slot.

Unlike unslotted ALOHA, the duration of a collision interval is now constant (Figure 5.3). The vulnerable period is L or $(L + \tau)$ if a propagation delay of τ is included. However, the average delay performance of slotted ALOHA is slightly worse than unslotted ALOHA since a packet cannot be transmitted immediately unless it arrives exactly at the slot boundary.

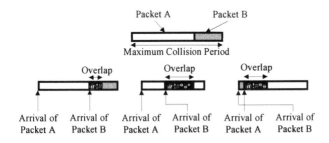

Figure 5.1: Two-packet collision scenarios in unslotted ALOHA

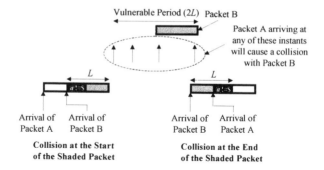

Figure 5.2: Vulnerable period in unslotted ALOHA

It is interesting to compare the conditions for collision in unslotted and slotted ALOHA. A collision in unslotted ALOHA occurs when one or more packets arrive before the previous packet has completed transmission. For slotted ALOHA, a collision occurs when two or more packets arrive within the same time slot. This is illustrated in Figure 5.3 where the interarrival times between packets 1 and 2, and between packets 3 and 4 are chosen to be the same. A collision occurs only between packets 3 and 4 for slotted ALOHA but in the case of unslotted ALOHA, all packets collide.

Note that whenever packets collide in slotted ALOHA, changing the protocol from slotted to unslotted does not remove the collision. However, the reverse is not true. In other words, whenever packets experience a collision in unslotted ALOHA, changing the protocol from unslotted to slotted may remove the collision. This suggests that more packets collide in unslotted ALOHA than slotted ALOHA. The throughput analysis that follows justifies this observation.

5.2.3 Disciplined ALOHA

In slotted ALOHA, users receive packets regardless of who the packet is intended for. When a new packet arrives at a user, the ongoing reception of any packet not destined for the user is immediately aborted. This results in performance degradation since if the user is locked on to a packet not destined for it, then it is very likely that there exists some other user which the packet is destined. Thus, it may advisable to inhibit transmission. Disciplined ALOHA solves this problem by allowing a user to transmit only when it is not already receiving any packet [TOBA87].

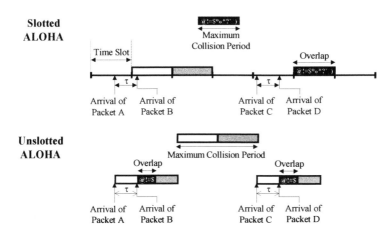

Figure 5.3: Two-packet collision scenarios in unslotted and slotted ALOHA

5.2.4 Spread ALOHA

In unslotted ALOHA, packets that overlap even by a small amount result in two or more corrupted packets that must be retransmitted over the channel. Spread spectrum receivers, however, are capable of compressing the received packets in such a way that the total period of packet overlap at the output of the receiver is much less than the period of overlap over the physical channel. The purpose of spreading packet transmission time over the channel is to lower the transmit power requirements while keeping the transmitted energy per bit constant [ABRA94]. The performance of spread ALOHA relies heavily on the time capture property of direct sequence spread spectrum systems (see Section 6.1).

5.3 PERFORMANCE ANALYSIS OF ALOHA PROTOCOLS

Before carrying out some fundamental analysis on the performance of ALOHA protocols, it is appropriate to define several symbols to be used in the analysis (Table 5.1).

5.3.1 Throughput Analysis

Consider an infinite population of users, each generating fixed-length packets under Poisson arrivals. The maximum throughput can be obtained by assuming steady-state (statistical) equilibrium with aggregate arrival process (from new and retransmitted packets) forming a Poisson distribution.

For unslotted ALOHA, the vulnerable period is $2T$. This gives:

$$P_0 = e^{-\frac{G \times 2T}{T}} = e^{-2G}$$

(5.1)

$$S = G \times P_0 = Ge^{-2G}$$

(5.2)

$$\frac{dS}{dG} = e^{-2G} - 2Ge^{-2G}$$

(5.3)

$$\frac{dS}{dG} = 0 \Rightarrow G = 0.5$$

(5.4)

Table 5.1: Definitions for performance analysis

Symbol	Definition
T	Average packet transmission time
S	Channel throughput (i.e., the average number of successful transmissions per packet transmission time, T)
G	Offered load (i.e., the average number of attempted transmissions per packet transmission time, T)
E	Average number of retransmissions
P_0	Probability of success (i.e., probability of no packets generated during vulnerable period)

$$\therefore S_{max} = \frac{1}{2e} = 0.184$$

(5.5)

For slotted ALOHA, the vulnerable period is T. This gives:

$$P_0 = e^{-\frac{G \times T}{T}} = e^{-G}$$

(5.6)

$$S = G \times P_0 = Ge^{-G}$$

(5.7)

$$\frac{dS}{dG} = e^{-G} - Ge^{-G}$$

(5.8)

$$\frac{dS}{dG} = 0 \Rightarrow G = 1$$

(5.9)

$$\therefore S_{max} = \frac{1}{e} = 0.368$$

(5.10)

Figure 5.4 illustrates the throughput performance of both protocols. The throughput of slotted ALOHA is maximized when the total channel traffic or offered load (i.e., new plus retransmitted packets) is unity (i.e., one packet per packet transmission time). Since the maximum throughput of the slotted ALOHA protocol is $1/e$ or 0.368, this means that on the average, each packet must be transmitted e (2.718) or roughly three times. Thus, less than 40% of a packet per packet transmission time can be transmitted successfully in ALOHA channels.

The maximum achievable throughput for unslotted ALOHA has been shown to be less than the $1/2e$ value when packets with variable lengths are sent [FERG77], [BELL80]. Although the slotted version of ALOHA has the advantage of throughput efficiency, this is offset by the need for synchronization and the increased header overhead when long packets are segmented into shorter packets that fit the duration of a fixed slot length. Furthermore, users have to delay their transmissions until the start of the slot boundary. Thus, the choice between unslotted and slotted ALOHA depends on the ratio of header overhead to packet lengths, the distribution of these lengths, the efficiency operating point chosen to satisfy delay and stability considerations, and the system cost of synchronization.

5.3.2 Average Number of Retransmissions

The ratio G/S is a measure of the average delay incurred in a ALOHA system since it represents the average number of transmissions before a packet is successfully transmitted. For slotted ALOHA, the probability of a transmission requiring exactly $k - 1$ attempts followed by one success is:

$$P_k = e^{-G}\left(1 - e^{-G}\right)^{k-1}$$

(5.11)

The average number of transmissions is:

$$Q = \sum_{k=1}^{\infty} kP_k = \sum_{k=1}^{\infty} ke^{-G}\left(1 - e^{-G}\right)^{k-1} = e^{G} \Rightarrow S = \frac{G}{e^{G}} = \frac{\ln Q}{Q}$$

(5.12)

Therefore, the average number of retransmissions is:

$$E = Q - 1 = e^{G} - 1$$

(5.13)

For unslotted ALOHA,

$$E = e^{2G} - 1$$

(5.14)

Figure 5.5 illustrates the number of retransmissions as a function of the channel throughput (S). Note that if Q is known, then for slotted ALOHA, S can be obtained directly from equation 5.12. Conversely, if S is given, then G can be obtained recursively from equation 5.7. Q is then computed from equation 5.12 using G.

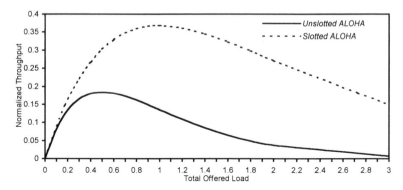

Figure 5.4: Throughput performance for unslotted and slotted ALOHA

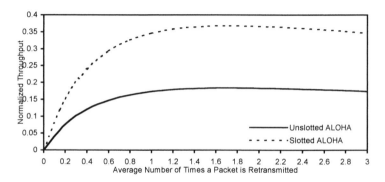

Figure 5.5: Number of retransmissions in unslotted and slotted ALOHA

5.3.3 Acknowledgments

Acknowledgments (ACKs) use part of the total channel bandwidth. In the analysis carried out so far, it has been assumed that ACKs are transmitted on a separate channel. When the available bandwidth is provided as a single channel to be shared by both data and ACK packets, then the channel performance will further suffer from interference between data and ACK packets unless some kind of priority scheme is provided.

It has been shown that in a common-channel configuration with non-prioritized ACK traffic, the maximum throughput of slotted ALOHA drops to 0.14 [TOBA89]. However, if there is some means where ACK traffic can be given priority so as to guarantee its transmission free of conflict, then the channel capacity for slotted ALOHA can be maintained at around 0.26 (assuming that an ACK packet occupies an entire slot). For example, ACKs of correctly received packets can be broadcast by a central controller to all users over a side channel which can be made very reliable because of its extremely low data rate requirement [RAMA87].

5.3.4 Power Capture

In wireless networks, it is possible for a radio receiver to "capture" the packet with the strongest signal power in spite the presence of weaker signals generated by other packets. Power capture is mainly due to the discrepancy in instantaneous power among received signals from users in different locations. This discrepancy can be attributed to distance, transmitter power, and fading effects.

For perfect capture, one of the packets survives each collision and the maximum throughput is obtained with maximum retransmission probability. This is in contrast to channels with no capture where all the packets involved in collisions are destroyed and the maximum throughput is associated with small values of the retransmission probability.

Power capture can improve the overall network performance. By means of adaptive power control, it can achieve either fairness to all users or intentional discrimination. With capture, the throughput of a network is better in fading than in nonfading channels (where the received power from all users is the same). For instance, in Rayleigh fading channels, the 0.368 maximum throughput of slotted ALOHA increases to about 0.7 [PAHL94].

5.3.5 Analyzing the Slotted ALOHA Protocol with Capture

Suppose a group of contending users is subdivided into two groups of transmit power levels – high and low. Note that high-power packet transmissions are not affected by low-power packet transmissions [SAAD94]. However, the converse is not true. Let S_H (S_L) be the average number of successful high-power (low-power) packet transmission per time slot and G_H (G_L) be the average number of high-power (low-power) packet transmission (newly generated plus retransmitted packets) per time slot.

For Poisson arrivals,

$$S_H = G_H e^{-G_H}$$

(5.15)

For a low-power packet to be successfully transmitted, the following events must occur:

❑ no other low-power packet in the same slot (probability of e^{-G_L});
❑ no high-power packet in the slot (probability of e^{-G_H}).

Assuming both events are independent, the probability of successful transmission is $e^{-(G_H + G_L)}$.

$$S_L = G_L e^{-(G_H + G_L)}$$

(5.16)

$$S_{L,\max} = \frac{1}{e} e^{-G_H}$$

(5.17)

$$S_{total} = S_{L,\max} + S_H = \frac{1}{e} e^{-G_H} + G_H e^{-G_H}$$

(5.18)

$$\frac{dS_{total}}{dG_H} = -\frac{1}{e} e^{-G_H} + e^{-G_H} - G_H e^{-G_H}$$

(5.19)

$$\frac{dS_{total}}{dG_H} = 0 \Rightarrow G_H = 1 - \frac{1}{e}$$

(5.20)

$$S_{total,\max} = e^{-\left(1 - \frac{1}{e}\right)} = 0.53$$

(5.21)

5.3.6 Controlled ALOHA

Suppose there are N ready users in the network, each transmitting with a probability of p. Then for slotted ALOHA:

$$S = \binom{N}{1} p(1-p)^{N-1} = Np(1-p)^{N-1}$$

(5.22)

$$\frac{dS}{dp} = N(1-p)^{N-1} - N(N-1)p(1-p)^{N-2}$$

(5.23)

$$\frac{dS}{dp} = 0 \Rightarrow p = \frac{1}{N}$$

(5.24)

$$S_{max} = \left(1-\frac{1}{N}\right)^{N-1} \rightarrow \frac{1}{e} \text{ or } 0.368 \text{ as } N \rightarrow \infty$$

(5.25)

The plot of the maximum throughput against the number of ready users is shown in Figure 5.6. Clearly, for a small number of ready users, the probability of success is high. When the number is high, the throughput degenerates to the asymptotic value of $1/e$. Thus, asymmetry in the probability of transmission provides certain benefits. It is possible to prove that optimal multiple access strategies are always asymmetric [LAM84]. To exploit this characteristic, the number of contending users (N) must be somehow be split into smaller groups as this number increases. Such an adaptive strategy optimizes the throughput when each ready user transmits in the next time slot with a probability of $1/N$.

5.3.7 Asymmetric Traffic Load

One important characteristic of the ALOHA protocol is that a user can potentially use a large portion of the channel capacity for some arbitrary period of time, depending on the traffic load of other users during that time. The average throughput efficiency can exceed $1/2e$ or $1/e$ when the traffic loads of the users are unequal.

Consider the case of slotted ALOHA with one large user (offered load G_L) and N identical small users (offered load G_S). Suppose $G_L = mG_S$ where $m \geq 1$. The respective throughputs are:

$$S_L = G_L (1 - G_S)^N = mG_S (1 - G_S)^N$$
(5.26)

$$S_S = G_S (1 - G_S)^{N-1} (1 - G_L) = G_S (1 - G_S)^{N-1} (1 - mG_S)$$
(5.27)

The overall throughput is:

$$S = S_L + NS_S$$
(5.28)

The maximum throughput occurs when:

$$NG_S + G_L = 1$$
(5.29)

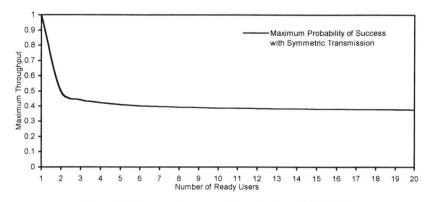

Figure 5.6: Symmetric transmission in slotted ALOHA

$$\therefore G_S = \frac{1}{m+N}$$

(5.30)

The maximum throughput becomes:

$$S_{max} = \left(\frac{m}{m+N}\right)\left(\frac{m+N-1}{m+N}\right)^N + \left(\frac{N}{m+N}\right)^2\left(\frac{m+N-1}{m+N}\right)^{N-1}$$

(5.31)

$$\therefore S_{max} = \frac{(m+N-1)^{N-1}}{(m+N)^{N+1}}\left(m^2 + mN - m + N^2\right)$$

(5.32)

The maximum throughput is plotted in Figure 5.7. Clearly, at high degrees of traffic load asymmetry, the throughput is much better than the symmetrical case. This result is known as the excess capacity of an ALOHA channel. It can achieve a throughput efficiency approaching unity in the case of a single user with a high packet rate and all other users with a very low packet rate. However, the delay encountered by the low-rate users is significantly higher than in the homogenous case [BIND81], [SCHW77].

5.3.8 Scheduling Retransmissions

When a collision occurs, each user involved is said to be backlogged. Since the objective is to minimize the average delay experienced by a packet between the time it was generated to the time it is successfully transmitted, this is equivalent to minimizing the average backlog in the network (by Little's Theorem). A backlogged user retransmits its packet after a randomized delay (random timeout). The randomly selected time reduces the likelihood of a collision from recurring (Figure 5.8). This randomized delay turns out to be crucial to the stability behavior and generally affects the throughput-delay performance of all contention-based protocols.

Two methods of scheduling retransmissions in slotted ALOHA following the detection of a collision have been considered:

❏ Geometrically distributed delay – a collided packet is retransmitted in a time slot with probability $p < 1$. With probability $1 - p$, the packet is delayed. The Bernoulli trial is repeated in the next time slot with a smaller value of p. A small p implies large delay between retrials of a collided packet;

❏ Uniformly distributed delay – a time slot is selected for retransmitting a collided packet in the next K time slots, each chosen with equal probability.

Consider the first method of retransmission. If a collision occurs in the nth slot, all users contend for the subsequent slot with a probability of:

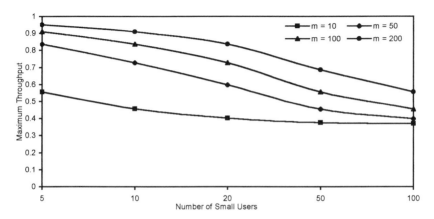

Figure 5.7: Slotted ALOHA with asymmetric loads

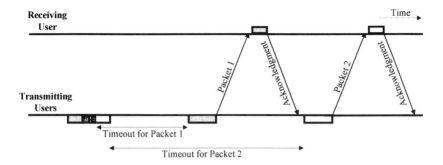

Figure 5.8: Retransmission in ALOHA protocols

$$p_{n+1} = \frac{1}{2} p_n$$

$$(5.33)$$

where $n = 1, 2, \ldots$

If A_n represents the probability of n transmissions, then:

$$A_1 = \binom{N}{1} p(1-p)^{N-1} = Np(1-p)^{N-1}$$

$$(5.34)$$

$$A_2 = (1 - A_1) N \frac{p}{2} \left(1 - \frac{p}{2}\right)^{N-1}$$

$$(5.35)$$

$$A_n = (1 - A_1 - A_2 \ldots - A_{n-1}) \frac{Np}{2^{n-1}} \left(1 - \frac{p}{2^{n-1}}\right)^{N-1} = \left(1 - \sum_{i=1}^{n-1} A_i\right) \frac{Np}{2^{n-1}} \left(1 - \frac{p}{2^{n-1}}\right)^{N-1}$$

$$(5.36)$$

The average number of slots elapsing before a successful packet transmission occurs is:

$$\overline{W} = \sum_{n=1}^{\infty} n A_n$$

$$(5.37)$$

Consider the second retransmission strategy where the packet is retransmitted in the next K slots, each selected with probability $1/K$. The average number of slots a packet has to wait before retransmission is:

$$\overline{W} = T \sum_{n=0}^{K-1} n \times \text{probability of selecting slot } n = T \sum_{n=0}^{K-1} n \times \frac{1}{K} = \frac{K-1}{2} T$$

$$(5.38)$$

Simulation studies indicate that in many cases, it is the average value and not the exact probability distribution of the retransmission delay is sufficient for predicting the behavior of slotted ALOHA [LAM74].

5.3.9 Delay Analysis

As mentioned earlier, a collided packet must be rescheduled in such a way that the retransmitted traffic is spread out and the likelihood of repeated collisions is reduced. Suppose the packet is retransmitted in the next K slots, each selected with probability $1/K$. The average number of slots a packet has to wait before retransmission is given in equation 5.38.

The average waiting time between the arrival of a packet and beginning of the following slot is half a slot. Taking into account the delay incurred during collisions, the total average delay experienced by a packet from the moment it arrives to time it is successfully transmitted can be calculated.

For unslotted ALOHA,

$$\overline{D} = (\tau + T + \overline{W})E + (\tau + T) = \left(\tau + \frac{K+1}{2}T\right)\left(e^{2G} - 1\right) + (\tau + T)$$

$$(5.39)$$

If the propagation delay τ is ignored, then the total average delay can be simplified to:

$$\overline{D} = T + \left(\frac{K+1}{2}\right)\left(e^{2G} - 1\right)T$$

$$(5.40)$$

For slotted ALOHA, packets cannot transmit instantly but must be delayed until the next slot. This introduces an average delay of half a time slot:

$$\overline{D} = \frac{1}{2}T + (\tau + T + \overline{W})E + (\tau + T) = \frac{1}{2}T + \left(\tau + \frac{1}{2}T + \frac{K+1}{2}T\right)\left(e^{G} - 1\right) + (\tau + T)$$

$$(5.41)$$

If the propagation delay is ignored, then:

$$\overline{D} = \frac{3}{2}T + \left(\frac{K+1}{2}\right)\left(e^{G} - 1\right)T$$

$$(5.42)$$

Figure 5.9 shows the delay performance of unslotted and slotted ALOHA for $K = 6$. The delay has been normalized with respect to the packet transmission time (T).

5.4 STABILITY PROBLEMS IN ALOHA PROTOCOLS

In Section 5.3, the maximum throughput of unslotted and slotted ALOHA is computed under conditions of statistical equilibrium. This assumes that the protocol will reach steady state equilibrium where traffic from retransmission will form a stationary Poisson process that is independent of the packet arrival rate. It is important to note that the maximum throughput is potentially unstable and therefore, unachievable in practice unless some form of adaptive control is applied. The notion of instability is also clear from the throughput curves of Figure 5.4 which suggests that any throughput can be achieved with two different values of the offered load.

As an example, suppose a slotted ALOHA system is observed over a long period of time. It is found that 25% of the slots contain successful transmissions. The offered load (G) can then be computed from the equation $0.25 = Ge^{-G}$, giving $G = 0.357$ and $G = 2.15$. This implies that for a packet transmission time (T) of 1 ms, the overall packet arrival rate (λ) is 357 packets/s and 2150 packets/s respectively. Note that the slotted ALOHA system is stable when $G < 1$ and unstable (overloaded) for $G > 1$.

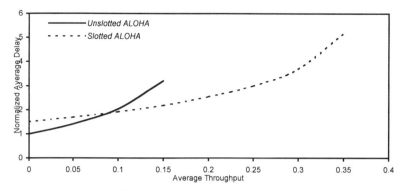

Figure 5.9: Delay performance for unslotted and slotted ALOHA $(K = 6)$

It is interesting to compare the percentage of empty and collision slots in both systems.

For a stable system,
Percentage of empty slots = e^{-G} = $e^{-0.357}$ = 0.7
Percentage of collision slots = $1 - e^{-G} - Ge^{-G}$ = $1 - 0.25 - 0.7 = 0.05$

For an unstable system,
Percentage of empty slots = $e^{-2.15}$ = 0.116
Percentage of collision slots = $1 - 0.25 - 0.116 = 0.634$

ALOHA networks tend to break down completely if they are exposed to an overload, even if this overload lasted only very shortly. Furthermore, even when the offered traffic is below capacity, such networks will eventually drift into a saturation state where the channel is filled with retransmissions (from backlogged users) and very low or zero throughput results. Once this state is reached, the system cannot return to its normal mode of operation without external intervention (e.g., resetting the entire system). Thus, although the throughput may be satisfactory for a short period of time, it becomes poor when observed over a long period of time.

5.4.1 Stability Characterization

A Markovian chain formulation of the infinite-user slotted ALOHA model shows that it is always unstable because the stationary probability distribution does not exist [KLEI75a], [TOBA89]. This instability is due to statistical fluctuations in the offered traffic load which increases the number of collisions. This in turn increases the offered load which then increases the frequency of collisions. Such positive feedback causes the throughput to decrease to very low values. Thus, a more accurate measure of channel performance must reflect the tradeoffs among stability, throughput and delay. This means that to adequately specify the performance of an ALOHA channel without an explicit stability control mechanism, a probability of expected stable operation time must be provided along with values of throughput efficiency and delay. Generally, this probability is a function of the interval used for retransmissions and the total number of users, with longer intervals resulting in a higher probability of maintaining stable operation for a given time.

Fortunately, the number of users, N, must be finite in a real network. In this case, the slotted ALOHA channels may exhibit stable or bistable behavior depending on the parameters of N, K (average retransmission time) and λ (input channel rate). Bistable systems operate with equilibrium points that can be desirable or undesirable. A desirable point is one where the throughput is high and very few users are backlogged. On the other hand, an undesirable point carries very little throughput even though most of the users are backlogged and the load is high. A bistable system that starts from the desirable operating point will, after a sufficiently long time, degenerate to the undesirable point with non-zero probability.

5.4.2 Dynamic Controls

Control algorithms are clearly needed for preventing saturation in ALOHA channels and guarantee that a successful transmission occurs within a finite number of slots following a collision. These mechanisms are usually based on one or more of the following [LAM83]:

❑ reducing the probability of retransmitting a collided packet in a time slot (thus increasing the backoff factor K);
❑ revoking the access rights of some users for a period of time (thus reducing N).

The goal of most adaptive control algorithms for slotted ALOHA is to achieve the unity offered load condition (i.e., $G = 1$). The main difficulty to achieving this condition is that the total number of ready users is usually unknown to each user.

5.5 FEEDBACK ALGORITHMS

Several proposals to overcome the instability of the slotted ALOHA protocol require the use of channel feedback based on the history of past transmissions. The central idea behind these sequential detection schemes is to control the retransmission process when resolving collisions. This is achieved by probabilistically splitting the set of packets involved in a collision into a transmitting set and a nontransmitting set while making all other packets not involved in the collision wait. The "collision resolution" approach opened the possibility of an exact analysis of the steady state behavior of a random access system. It eliminates the statistical equilibrium hypothesis commonly employed in the analysis of ALOHA systems.

5.5.1 The Binary Tree Algorithm

One of the first collision resolution proposals was the binary tree algorithm [CAPE77], [CAPE79a] which can provide a maximum capacity of 0.43. The binary tree algorithm is closely related to the probing method described in Section 4.5.1 with the exception that no central controller is necessary. Thus, it is based on the observation that contention among several ready users is completely resolved if the users are successively divided into smaller groups. Each group is restricted to transmit on a separate time slot. This process continues until a single contending user is isolated in each group. The depth of the tree (i.e., the number of times the contending users have to be divided) increases as the number of ready users increases.

In the tree algorithm, the order of transmission is not taken into account. Following a collision, all users with addresses starting with 0 transmit if they had packets involved in the collision. On the next slot, users with addresses starting with 1 transmit. If conflicts occur in either of these slots, the corresponding set of users is further partitioned into smaller sets specified by the first two bits of the address and so forth until the packets are successfully transmitted (Figure 4.5).

5.5.2 Analyzing the Tree Algorithm

Suppose two users employ the tree algorithm for channel access. The collision resolution period is minimum (one slot) when only one user transmits or when no user transmits. The resolution period becomes maximum (three slots) when both users transmit. Note that the collision resolution period cannot be two slots. Assume that the two users transmit new packets in a slot with probability P_A and P_B, and that user A has address 0 while user B has address 1. This implies that when a collision occurs in the first slot, user A will retransmit in the second slot while user B will retransmit in the third slot.

The probability that user A transmits and user B does not is:

$$\alpha_1 = P_A(1 - P_B)$$

$$(5.43)$$

Similarly, the probability that user B transmits and user A does not is:

$$\alpha_2 = P_B(1 - P_A)$$

(5.44)

The probability that one user transmits while the other does not is:

$$\sigma_1 = \alpha_1 + \alpha_2 = P_A(1 - P_B) + P_B(1 - P_A)$$

(5.45)

The probability that no user transmits is:

$$\sigma_0 = (1 - P_A)(1 - P_B)$$

(5.46)

The probability that two users transmit is:

$$\sigma_2 = P_A P_B$$

(5.47)

The probability that one user transmits while the other does not is:

$$F = 1 \times (\sigma_0 + \sigma_1) + 3 \times \sigma_2 = 1 + P_A P_B$$

(5.48)

The average throughput is the ratio of successful slots to the total number of slots in a collision resolution period and is given by:

$$S = \frac{1}{1}\sigma_1 + \frac{2}{3}\sigma_2 = P_A + P_B - \frac{4}{3}P_A P_B$$

(5.49)

The average delay for user A is:

$$D_A = (1 \times \alpha_A) + (2 \times \sigma_2) = P_A + P_A P_B$$

(5.50)

The average delay for user B is:

$$D_B = (1 \times \alpha_B) + (3 \times \sigma_2) = P_A + 2P_A P_B > D_A$$

(5.51)

Clearly, a user with a higher numbered address (e.g., address 1) will suffer a higher delay than a user with a lower numbered address (e.g., address 0).

Suppose $P_A = P_B = P$. Then from equations 5.46 to 5.49:

$$F = 1 + 2P^2$$

(5.52)

The average delay for user A is:

$$D_A = P + P^2$$

(5.53)

$$D_B = P + 2P^2$$

(5.54)

$$S = 2P - \frac{4}{3}P^2$$

(5.55)

$$\frac{dS}{dP} = 2 - \frac{8}{3}P = 0 \Rightarrow P = 0.75$$

(5.56)

$$\therefore S_{max} = 0.75$$

(5.57)

The variation of F, D_A, D_B, and S_{max} with respect to P is illustrated in Figures 5.10 and 5.11. Clearly, the throughput performance for the tree algorithm is superior to slotted ALOHA.

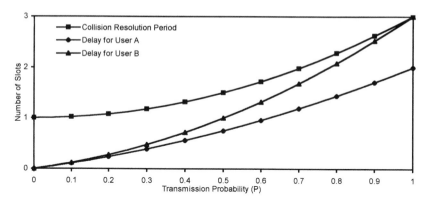

Figure 5.10: Delay performance and collision resolution period

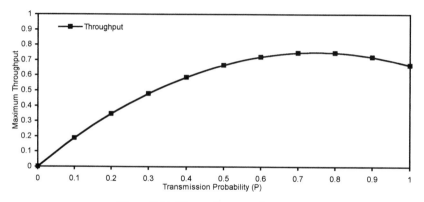

Figure 5.11: Throughput performance

5.5.3 Improving the Tree Algorithm

Since new packets are delayed until all conflicts are resolved, if the tree algorithm spends many slots resolving a collision, then many new packets typically accumulate. When the collision resolution period terminates, these waiting packets will cause a definite collision and many successive collisions will follow. This situation can be improved by creating a new resolution period immediately after the end of the previous period and then randomly select a waiting packet to join one of k subgroups. In order to reduce the likelihood of multiple-packet collisions, the number k increases with the length of the preceding collision resolution period.

Another improvement to the algorithm is to avoid a predictable collision by immediately resplitting the second subgroup when the first subgroup was found to be empty (Figure 5.12). With this modification, the maximum capacity is increased to 0.46. The improvement comes at a cost – when channel errors occur, the modified algorithm suffers from deadlock i.e., after some point, collisions are never resolved, and no packets can be transmitted [MASS81]. Nevertheless, unlike ALOHA, these schemes are stable if operated below their capacities.

5.5.4 The Splitting Algorithm

The binary tree algorithm can be improved upon using the splitting algorithm [GALL78] that schedules retransmission by dividing contending users into smaller subgroups based on the time of arrival of their packets. The improvement recognizes the fact that packets are generated at unique instants on the time axis. The splitting algorithm identifies the packet generation times as follows.

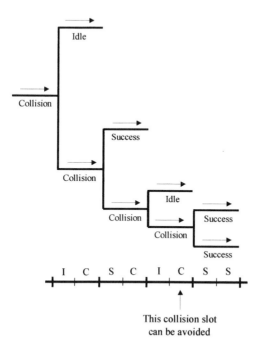

Figure 5.12: Improving the binary tree algorithm

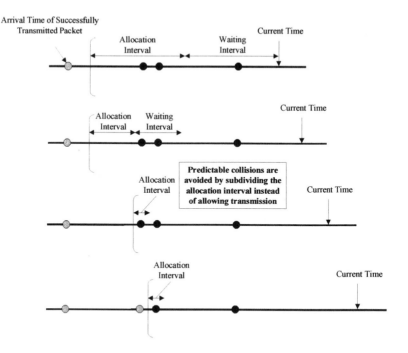

Figure 5.13: Splitting algorithm

Each subgroup consists of packets which have arrived within a specified time interval. If there is a collision, this interval is split into two equal subintervals (Figure 5.13). The earlier arriving packet is then allowed to transmit first (in the first subinterval) and therefore, packets will be transmitted in the order of arrival (i.e., on a first-come-first-served basis). Thus, the delay variance is smaller than random order. Note that for the protocol to function, all users must be aware of a common time window in the past and only users with a packet generated during the window are allowed to transmit.

5.5.5 Implementation Considerations

The implementation of both the binary tree and splitting algorithms is distributed, requiring users to be able to observe outcomes of the broadcast channel continuously and make decisions. To do so, the channel needs to be time slotted and the user must be capable of determining after transmission of a packet whether there has been zero (i.e., idle channel), one (i.e., success), or multiple packets (i.e., collision). Furthermore, each algorithm step requires at least a channel propagation time plus two time slots to execute. Like probing, tree search protocols suffer in performance when there are many ready users.

5.6 CARRIER SENSE MULTIPLE ACCESS

For channels with short propagation delay compared to the packet transmission time (e.g., indoor wireless networks), collisions can be significantly reduced by requiring each ready user to sense the channel for the presence of any ongoing transmission before transmitting (i.e., "listen before talk"). In this case, when a user is sending a packet, all other users in the network become aware of the transmission within a fraction of the packet transmission time. Such protocols are also known as deference protocols because during the packet transmission period, all other users must remain silent (i.e., all users must defer to an ongoing transmission).

Well-known deference protocols that implement carrier sensing include the Carrier Sense Multiple Access (CSMA) family of protocols. With CSMA, the vulnerable collision period is now reduced to twice the maximum propagation delay, τ. If the channel is divided into minislots of duration τ, the vulnerable period is reduced even further to just τ seconds. In addition to reducing the number of collisions, the minislots also help to shorten the time for a new packet to access the channel since the slot boundaries now arrive more regularly than in slotted ALOHA.

5.6.1 CSMA Variations

In ALOHA, the action taken by the users is binary (i.e., to transmit or not to transmit). In CSMA, several transmission strategies are possible. Two main CSMA protocols are known as non-persistent and p-persistent CSMA. These versions of CSMA differ according to the action taken by a ready user after sensing an idle channel. However, when a transmission is unsuccessful, each protocol schedules the retransmission of the packet in the same way (i.e., retransmitting after a random delay).

In the non-persistent protocol, the ready user listens the channel and operates as follows:

❑ if the channel is sensed idle, the packet is transmitted immediately;
❑ if the channel is sensed busy, the user waits for a random amount of time before resensing the channel.

The p-persistent protocol is used for time-slotted channels only. In this case, the ready user listens to the channel and operates as follows:

❑ if the channel is sensed idle, the user transmits with a probability of p (or defers until the next slot with a probability of $1 - p$);
❑ if the channel is sensed busy, the user continues to listen until the channel goes idle and then transmits with a probability of p (or defers until the next slot with a probability of $1 - p$.

A special case of p-persistent CSMA is 1-persistent CSMA, which allows a packet to be transmitted immediately when the channel is sensed to be free. It was designed to achieve a higher throughput than p-persistent CSMA since the channel is never allowed to go idle whenever users have packets to transmit. However, this is done at the expense of a higher collision probability, particularly after the end of a packet transmission where new packets tend to accumulate. In all cases, due to a finite propagation delay, it is possible for a user to sense a channel as idle when another user has just begun transmitting, thus leading to a collision (Figure 5.14).

5.6.2 Performance Considerations

Suppose τ is the maximum one propagation delay and T is the packet transmission time. The ratio $a = \tau/T$ has a profound effect on the maximum channel throughput of a CSMA protocol. Specifically, the maximum throughput of CSMA approaches unity if a is decreased to zero. With zero propagation delay, a becomes zero and collisions in CSMA are completely avoided. The protocol's performance is then equivalent to an M/D/1 queue. In this case, the cost of creating a global queue disappears.

The maximum channel throughput of CSMA can be obtained as a function of the equilibrium slotted ALOHA throughput (S_{SA}) [LAM80]:

$$S_{CSMA} = \frac{S_{SA}}{2a + S_{SA}(1 + a)}$$

$$(5.58)$$

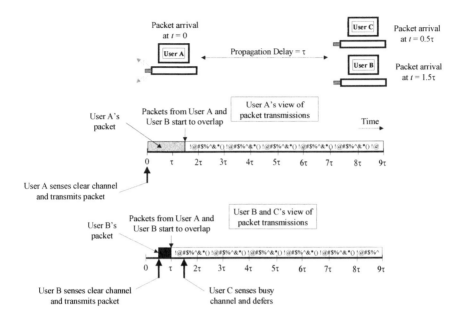

Figure 5.14: Operation of CSMA

For an indoor wireless environment, *a* is small (typically 0.001 to 0.1) since τ is short. This means that when the channel is lightly loaded, the delay in identifying a ready user and assigning channel access using CSMA is extremely short. This delay is independent of the number of users (unlike polling or probing protocols). In particular, when there is only one user, the delay is zero.

Like all contention schemes, the channel throughput for CSMA becomes inferior when the network is heavily loaded. However, if packets can be queued, then the overhead per packet is reduced. In the limit of infinitely long queues at each user, the channel throughput of CSMA approaches unity. This situation is similar to the case of ALOHA with power capture.

5.6.3 Implementation Considerations

Like ALOHA, the weakness of CSMA is that channel access has to be negotiated for each packet. Carrier sensing reduces transmission efficiency because the turnaround time in radio transceivers can take up a nontrivial portion of a packet's transmission time [BING00]. Furthermore, the advantage of carrier sensing is lost in a partially-connected network where hidden users (i.e. users within range of the intended destination but not of the transmitter) can severely degrade the throughput of CSMA. For example, a collision can occur at the receiver even though the transmitter sensed the channel as being idle. In Figure 5.15, user A transmits to user B. At the same time, if user C transmits to user D (or user B), a collision will occur at user B. The problem arises because user C is out of the range of user A, and is therefore unable to sense the transmission from user A.

5.6.4 CSMA with Collision Detection

The performance of carrier sensing can be further improved by allowing users involved in a collision to abort their transmissions upon detecting the collision. Collided packets are retransmitted after a random delay and each collision involving the same packet doubles the retransmission delay. The reason for this improvement is that collision periods are shortened and collisions will not continue for the entire packet duration. Typical mechanisms for collision detection are based on comparing transmitted and received signals. This method is easily implemented in wired networks since the signal strength is essentially the same for a transmitter and receiver placed apart at a reasonable length. However, it appears that collision detection using this method is not feasible in a wireless environment due to the large differences (dynamic range) in signal strength, thus making any comparison difficult. In fact, a transmitted signal will easily overpower the received signals from all other users.

The "comb" scheme [HALL94] overcomes this limitation. Each user selects a random binary number which is employed as a key for detecting collisions. For each binary one in the key, a short burst of energy is transmitted. For each zero, the user must switch to the receive mode. Any user which detects the transmission of another (when operating in the receive mode) stops transmitting immediately. There is small probability of two users choosing the same key. This probability increases with the number of users in the network and decreases with the length of the key.

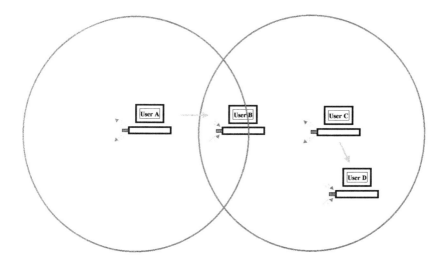

Figure 5.15: Hidden-user problem in CSMA

The major disadvantage of this approach is the fixed overhead required on every packet. The users must also be able to switch rapidly from the transmit to receive modes. An idea close related to the comb scheme is resource auction multiple access (RAMA) [AMIT92]. Here, users transmit simultaneously in a fashion similar to the comb scheme after hearing channel resources that are auctioned by the base station.

5.6.5 Virtual-Time CSMA

The use of two clocks in CSMA can provide the benefit of fairness based on arrival time. Like the splitting algorithm, it exploits the observation that packets arrive at unique time instants. The operation of virtual-time CSMA [MOLL85] requires a real-time clock and a virtual-time clock at each user. These two clocks advance in discrete units of slots. The real-time clock advances continuously while the virtual-time clock advances only when the channel is idle. When advancing, the virtual-time clock is incremented at two different rates relative to the real-time clock as follows:

❑ When virtual time lags behind real time, the virtual-time clock is advanced at a rate n times the real-time clock rate where $n > 1$;
❑ When virtual time has caught up with real time, the virtual clock rate is advanced at the same rate as real time or, equivalently, at a rate $n = 1$.

If a message arrives when virtual time has caught up with real time, the message is transmitted immediately at the next instant. On the other hand, if virtual time lags real time when a message arrives, the message will be marked with the current real-time value and held until virtual time catches up with this value. Note that for this scheme, two packets will collide only if their packet generation times are within the propagation delay of the network.

5.6.6 Busy Tone Multiple Access

One way of solving the hidden user problem in CSMA is the use of a busy tone, thus giving rise to busy tone multiple access (BTMA) [TOBA74]. The operation of BTMA depends on a central controller being able to sense the transmissions of all users and transmit the busy tone on a separate channel whenever transmissions are detected.

5.7 ADAPTIVE PROTOCOLS

Although network characteristics such as propagation delay and data rate are unlikely to vary during operation, traffic load patterns are likely to change over time. Thus, it will be useful if the access protocol is adaptive to changing traffic loads so that optimal channel utilization is maintained at all times. Some examples of adaptive access schemes include the probing, binary tree, and URN [KLEI78] protocols. The probing and binary tree methods have been discussed in Sections 4.5.1 and 5.5.1 respectively.

5.7.1 The URN Protocol

In the URN scheme, ready users decide whether or not to transmit in the next time slot with a probability of either 1 or 0. Thus, users have asymmetric access rights, some full while others none. This is in contrast to slotted ALOHA where all users are given the full access rights. It is also different from the controlled slotted ALOHA scheme (Section 5.3.6) where all users are given the same partial access right (i.e., the right to transmit with probability of $1/N$).

The URN scheme gives full access right to some subset of k users. A successful transmission results if there is exactly one ready user among the k users. Consider an analogy where each user is represented by a colored ball in an urn: black for ready and white for not ready. Let k be the number of balls sampled (drawn) from the urn. The probability of a successful transmission (S) is that of getting exactly one black ball in the sample. If N is the number of balls and n is the number of black balls, the probability is:

$$S = \frac{\binom{n}{1}\binom{N-n}{k-1}}{\binom{N}{k}}$$

(5.59)

This equation is maximized when $k = \lfloor N/n \rfloor$, k being a truncated integer. When this optimum value of k is chosen, it can be shown that the expected number of ready users selected (which is equal to the sample size k multiplied by the probability that a given ball is black) will be close one (not exactly one due to the truncation). Thus, not only does the value of k maximize the probability of selecting exactly one black ball (thereby maximizing the probability of a successful transmission), it also ensures that the average number of black balls selected is equal to one (i.e., $G = 1$).

To summarize, when n out of N users are busy, the URN scheme selects $k = \lfloor N/n \rfloor$ users. As the network load increases, n increases and the number of users given access rights (k) is reduced. For example, if $N = 12$ and $n = 1$, then k becomes 12 which implies that every user is given access rights (as in ALOHA). At the other extreme, when $n = 8$, $k = 1$. In this case, exactly one user will be given full access rights. Thus, there will be no collisions and the protocol behaves like TDMA.

5.8 HYBRID ACCESS PROTOCOLS

Hybrid access strategies integrate different access schemes. Some examples include split reservation upon collision, mixed ALOHA carrier sense, and random access polling. While these protocols attempt to combine the advantages of random and fixed access, they also suffer the combined drawbacks and overhead of both classes of access schemes.

5.8.1 Split Reservation Upon Collision

In SRUC [BORG78], all users operate in the random access mode under normal conditions. However, upon detection of collisions, the users will switch to the TDMA-reservation mode until all outstanding reservations are cleared. Thus, if a collision is detected, reservations are automatically implied for the colliding users. Chapter 7 covers reservation protocols in detail.

5.8.2 Mixed ALOHA Carrier Sense

In slotted ALOHA, more than 60% of the channel capacity is wasted. The MACS scheme allows a heavily loaded user to steal, by carrier sensing, slots which are unused by a large number of small users accessing the channel in a slotted ALOHA mode. Analysis has shown that the total channel utilization is significantly increased with MACS, and that the throughput-delay performance of both the large user and the background ALOHA users is better with MACS than with a split-channel configuration in which the larger user and the ALOHA users are each permanently assigned a portion of the channel [SCHO79].

5.8.3 Random Access Polling

In random address polling [CHEN94], the identities of all users in the network need not be known. The protocol essentially comprises two steps:

❑ A central controller first broadcasts a special signal that solicits responses from ready users;

❑ Ready users generate a random address and then send their requests simultaneously to the controller.

Since these transmissions are based on orthogonal signaling (e.g., using PN codes), the controller can decode the addresses simultaneously. The controller then polls the ready users.

SUMMARY

The ALOHA and CSMA family of protocols are attractive for low to medium load conditions but suffer from stability problems at high load. Adaptive multiple access techniques attempt to address this weakness. In the tree algorithm, users involved in a collision are broken down into smaller groups. This process continues until there is only one ready user in the group. The URN protocol restricts the number of users accessing the channel according to the traffic load. At light load, a large number of users are given access rights and this number is gradually reduced as traffic load increases. Hybrid access protocols attempt to combine the advantages of random and fixed access.

BIBLIOGRAPHY

[ABRA92]   Abramson, N., "Fundamentals of Packet Multiple Access for Satellite Networks", *IEEE Journal on Selected Areas in Communications*, Vol. 10, No. 2, February 1992, pp. 309 – 316.

[ABRA93]   Abramson, N., *Multiple Access Communications*, IEEE Press, 1993.

[ABRA94]   Abramson, N., "Multiple Access in Wireless Digital Networks", *Proceedings of the IEEE*, Vol. 82, No. 9, September 1994, pp. 1360 – 1370.

[AGRA83]   Agrawala, A. and Tripathi, S., *Performance '83*, North-Holland Publishing Company, 1983.

[AMIT92]   Amitay, N., "Resource Auction Multiple Access (RAMA): Efficient Method for Fast Resource Assignment in Decentralized Wireless PCS", *Electronics Letters*, April 1992, pp. 799 – 801.

[APOS85]   Apostolopoulos, T and Protonotarios, E., "Queueing Analysis of Buffered Slotted Multiple Access Protocols", *Computer Communications*, Vol. 8, No. 1, February 1985, pp. 9 – 22.

[BELL80]   Bellini, S. and Borgonovo, F., "On the Throughput of an ALOHA Channel with Variable Length Packets", *IEEE Transactions on Communications*, Vol. COM-28, No. 11, July 1984, pp. 1932 – 1935.

[BERG84]   Berger, T., Mehravari, N., Towsley, D. and Wolf, J., "Random Multiple-Access Communication and Group Testing", *IEEE Transactions on Communications*, Vol. COM-32, No. 7, July 1984, pp. 769 – 779.

[BIND81]   Binder, R., "Packet Protocols for Broadcast Satellites", appearing in [KUO81], pp. 175 – 201.

[BING00]   Bing, B., *High-Speed Wireless ATM and LANs*, Artech House, 2000.

[BORG78]   Borgonovo, F. and Fratta, L., "SRUC: A Technique for Packet Transmission on Multiple Access Channels Reservation and TDMA Schemes", *Proceedings of the ICCC*, September 1978, pp. 601 – 607.

[CAPE77]   Capetanakis, J., *The Multiple Access Broadcast Channel: Protocol and Capacity Considerations*, Ph.D Dissertation, MIT, 1977.

[CAPE79a]   Capetanakis, J., "Tree Algorithms for Packet Broadcast Channels", *IEEE Transactions on Information Theory*, Vol. IT-25, No. 5, September 1979, pp. 505 – 515.

[CAPE79b]   Capetanakis, J., "Generalized TDMA: The Multi-accessing Tree Protocol", *IEEE Transactions on Communications*, Vol. COM-27, No. 10, October 1979, pp. 1476 – 1484.

[CARL75]   Carleial, A. and Hellman, M., "Bistable Behavior of ALOHA-Type Systems", *IEEE Transactions on Communications*, Vol. COM-23, April 1975, pp. 401 – 410.

[CHEN94]   Chen, K., "Medium Access Control of Wireless LANs for Mobile Computing", *IEEE Network Magazine*, Vol. 8, No. 5, September 1994, pp. 50 – 63.

[DAVI80]   Davis, D. and Gronemeyer, S., "Performance of Slotted ALOHA Random Access with Delay Capture and Randomized Time of Arrival", *IEEE Transactions on Communications*, Vol. COM-28, No. 5, May 1980, pp. 703 – 712.

[FERG77]   Ferguson, M., "An Approximate Analysis of Delay for Fixed and Variable Length Packets in an Unslotted ALOHA Channel", *IEEE Transactions on Communications*, Vol. COM-27, No. 7, July 1977, pp. 644 – 654.

[GALL78]   Gallager, R., "Conflict Resolution in Random Access Broadcast Networks", *Proceedings of the AFOSR Workshop on Communication Theory and Applications*, September 1978, pp. 74 – 76.

[GERA84]   Gerakoulis, D., Saadawi, T. and Schilling, D., "A Class of Tree Algorithms with Variable Message Length", *Proceedings of ACM SIGCOMM*, 1984, pp. 242 – 247.

[GERL77]   Gerla, M., "Closed Loop Stability Controls for S-ALOHA Satellite Communications", *Proceedings of the 5th Data Communication Symposium*, September 1977, pp. 2-10 – 2-19.

[HALL94]   Halls, G., "HIPERLAN: The High Performance Radio Local Area Network Standard", *Electronics and Communication Engineering Journal*, December 1994, pp. 289 – 296.

[HAMM86]   Hammond, J. and O'Reilly, P., *Performance Analysis of Local Computer Networks*, Addison Wesley, 1986.

[HEYM82]   Heyman, D. P., "An Analysis of the Carrier Sense Multiple Access Protocol", *The Bell System Technical Journal*, Vol. 61, No. 8, October 1982, pp. 2023 – 2051.

[HIDE85]   Takagi, H. and Kleinrock, L., "Throughput Analysis for Persistent CSMA Systems", *IEEE Transactions on Communications*, Vol. COM-33, No. 7, July 1985, pp. 627 – 638.

[HLUC81]   Hluchyj, M. and Gallager, R., "Multiaccess of a Slotted Channel by Finitely Many Users", *Proceedings of the National Telecommunications Conference*, December 1981, pp. D4.2.1 – D4.2.7.

[IEEE85]   "Special Issue on Random Access Communications", *IEEE Transactions on Information Theory*, Vol. IT -31, No. 2, March 1985.

[ITAI84]   Itai, A. and Rosberg, Z., "A Golden Ratio Control Policy for a Multiple-Access Channel", *IEEE Transactions on Automatic Control*, Vol. AC-29, No. 8, August 1984, pp. 712 – 719.

[JENQ86]   Jenq, Y., "Theoretical Analysis of Slotted ALOHA, CSMA and CSMA-CD Protocols", appearing in BLAK86], pp. 325 – 346.

[KLEI75a]   Kleinrock, L., and Lam, S., "Packet Switching in a Multiaccess Broadcast Channel: Performance Evaluation", *IEEE Transactions on Communications*, Vol. COM-23, April 1975, pp. 410 – 423.

[KLEI75b]   Kleinrock, L. and Tobagi, F., "Packet Switching in Radio Channels: Part I", *IEEE Transactions on Communications*, Vol. COM-23, December 1975, pp. 1400 – 1416.

[KLEI78]   Kleinrock, L. and Yemini, Y., "An Optimal Adaptive Scheme for Multiple Access Broadcast Communication", *Proceedings of the ICC*, June 1978, appearing in [LAM84], pp. 195 – 199.

[KUMA84] Kumar, P. and Merakos, L., "Distributed Control of Broadcast Channels with Acknowledgment Feedback: Stability and Performance", *Proceedings of 23rd Conference on Decision and Control*, December 1984, pp. 1143 – 1147.

[KUO81] Kuo, F., *Protocols and Techniques for Data Communication Networks*, Prentice Hall, 1981.

[LAM74] Lam, S., *Packet Switching in a Multi-access Broadcast Channel with Applications to Satellite Communication in a Computer Network*, Ph.D Dissertation, UCLA, 1974.

[LAM75] Lam, S. and Kleinrock, L., "Packet Switching in a Multiaccess Broadcast Channel", *IEEE Transactions on Communications*, Vol. COM-23, September 1975, pp. 891 – 904.

[LAM80] Lam, S., "A Carrier Sense Multiple Access Protocol for Local Networks", *Computer Networks*, 1980, appearing in [LAM84], pp. 158 – 169.

[LAM83] Lam, S, "Multiple Access Protocols", appearing in [LAM84].

[LAM84] Lam, S., *Principles of Communication and Networking Protocols*, IEEE Computer Society Press, 1984.

[LONG81] Longo, G., *Multi-User Communications*, Springer-Verlag, 1981.

[LYNC87] Lynch, C and Brownrigg, E, *Packet Radio Networks: Architectures, Protocols, Technologies and Applications*, Pergamon Press, 1987.

[MARC83] Marcus, G., and Papantoni-Kazakos, P., "Dynamic Scheduling Protocols for a Multiple-Access Channel", *IEEE Transactions on Communications*, Vol. COM-31, No. 9, September 1983.

[MASS81] Massey, J., "Collision-Resolution Algorithms and Random Access Communications", appearing in [LONG81], pp. 73 – 137.

[MERA85] Merakos, L. and Kazakos, D., "On Retransmission Control Policies in Multiple-Access Communication Networks", *IEEE Transactions on Automatic Control*, Vol. AC-30, No. 2, February 1985, pp. 109 – 117.

[METC73] Metcalfe, R., "Steady State Analysis of a Slotted and Controlled ALOHA System with Blocking", *Proceedings of the 6th Hawaii International Conference on Systems Sciences*, January 1973, pp. 375 – 378.

[MITA81] Mittal, K. and Venetsanpoulous, A., "On the Dynamic Control of the Urn Scheme for Multiple Access Broadcast Communication Systems", *IEEE Transactions on Communications*, Vol. COM-29, No. 7, July 1981, pp. 962 – 970.

[MOLL83]   Molle, M., "A Simulation Study of Retransmission Strategies for the Asynchronous Virtual Time CSMA Protocol", appearing in [AGRA83].

[MOLL85]   Molle, M. and Kleinrock, L., "Virtual Time CSMA: Why Two Clocks are Better Than One", *IEEE Transactions on Communications*, Vol. COM-33, No. 9, June 1985, pp. 919 – 933.

[MOSE85]   Mosely, J. and Humblet, P., "A Class of Efficient Contention Resolution Algorithms for Multiple Access Channels", *IEEE Transactions on Communications*, Vol. COM-33, No. 2, February 1985, pp. 145 – 970.

[ONOZ80]   Onozato, Y., and Noguchi, S., "Dynamic Characteristics of a Satellite Communication System Employing the Slotted ALOHA Scheme", ", *Information Processing 80*, IFIP, North-Holland Publishing Company, 1980, pp. 581 – 585.

[PAHL94]   Pahlavan, K. and Levesque, A., "Wireless Data Communications", *Proceedings of the IEEE*, Vol. 82, No. 9, September 1994, pp. 1398 – 1430.

[PANW93]   Panwar, S., Towsley, D. and Armoni, Y., "Collision Resolution Algorithms for a Time-Constrained Multiaccess Channel", *IEEE Transactions on Communications*, Vol. 41, No. 7, July 1993, pp. 1023 – 1030.

[PAPA92]   Papantoni-Kazakos, P., "Multiple-Access Algorithms for a Systems with Mixed Traffic: High and Low Priority", *IEEE Transactions on Communications*, Vol. 40, No. 3, March 1992, pp. 541 – 548.

[RAMA87]   Ramamurthi, B., Saleh, A. and Goodman, D., "Perfect Capture for Local Radio Communications", *IEEE Journal of Selected Areas in Communications*, June 1987, appearing in [ABRA93].

[ROBE75]   Roberts, L., "ALOHA Packet System With and Without Slots and Capture", *Computer Communications Review*, Vol. 5, April 1975, pp. 28 – 42.

[ROSE83]   Rosenkrantz, W. and Towsley, D., "On the Instability of the Slotted ALOHA Multiaccess Algorithm", *IEEE Transactions on Automatic Control*, Vol. AC-28, No. 10, October 1983, pp. 994 – 996.

[ROSN82a] Rosner, R., *Packet Switching: Tomorrow's Communications Today*, Wadsworth, 1982.

[SAAD81]   Saadawi, T. and Ephremides, A., "Analysis, Stability and Optimization of Slotted ALOHA with a Finite Number of Buffered Users", *IEEE Transactions on Automatic Control*, Vol. AC-26, No. 3, June 1981, pp. 680 – 689.

[SAAD94]   Saadawi, T., Ammar, M. and Hakeem, A., *Fundamentals of Telecommunication Networks*, John Wiley, 1994.

[SANT80]   Sant, D., "Throughput of Unslotted ALOHA Channels with Arbitrary Packet Interarrival Time Distribution", *IEEE Transactions on Communications*, Vol. COM-28, No. 8, August 1980, pp. 1422 – 1426.

[SCHO79]   Scholl, M., "On a Mixed Mode Multiple Access Scheme for Packet-Switched Radio Channels", *IEEE Transactions on Communications*, Vol. COM-27, No. 6, June 1979, pp. 906 – 911.

[SCHW77]   Schwartz, M., *Computer Communications Network Design and Analysis*, Prentice Hall, 1977.

[SENN81]   Sennott, J. and Sennott, L., "A Queueing Model for Analysis of a Bursty Multiple-Access Communication Channel", *IEEE Transactions on Information Theory*, Vol. IT-27, No. 3, May 1991, pp. 317 – 321.

[SHAC83]   Shacham, N., "A Protocol for Preferred Access in Packet-Switching Radio Networks", *IEEE Transactions on Communications*, Vol. COM-31, No. 2, February 1983, pp. 253 – 264.

[SZPA88a]   Szpankowski, W., "Stability Conditions for Multidimensional Queueing Systems with Computer Applications", *Operations Research*, Vol. 36, No. 6, November/December 1988, pp. 944 – 954.

[SZPA88b]   Szpankowski, W. and Rego, V., "Some Theorems on Instability with Applications to Multiaccess Protocols", *Operations Research*, Vol. 36, No. 6, November/December 1988, pp. 958 – 967.

[TOBA74]   Tobagi, F., *Random Access Techniques for Data Transmission over Packet Switched Radio Networks*, Ph.D Dissertation, UCLA, 1974.

[TOBA75]   Tobagi, F. and Kleinrock, L., "Packet Switching in Radio Channels: Part II", *IEEE Transactions on Communications*, Vol. COM-23, December 1975, pp. 1417 – 1433.

[TOBA77]   Tobagi, F. and Kleinrock, L., "Packet Switching in Radio Channels: Part IV – Stability Considerations and Dynamic Control in Carrier Sense Multiple Access", *IEEE Transactions on Communications*, Vol. COM-25, No. 10., October 1977, pp. 1103 – 1119.

[TOBA78]   Tobagi, F. and Kleinrock, L., "The Effect of Acknowledgment Traffic on the Capacity of Packet-Switched Radio Channels", *IEEE Transactions on Communications*, Vol. COM-26, No. 6, June 1978, pp. 815 – 826.

[TOBA80]   Tobagi, F., "Analysis of a Two-Hop Centralized Packet Radio Network", *IEEE Transactions on Communications*, Vol. COM-28, No. 2, February 1980, pp. 196 – 205.

[TOBA82]   Tobagi, F., "Distributions of Packet Delay and Interdeparture Time in Slotted ALOHA and Carrier Sense Multiple Access", *Journal of the ACM*, Vol. 29, No. 4, October 1982, pp. 907 – 927.

[TOBA87]   Tobagi, F., "Modeling and Performance Analysis of Multihop Packet Radio Networks", *Proceedings of IEEE*, Vol. 75, No. 1, January 1987, pp. 135 – 155.

[TSIT87]   Tsitsiklis, J., "Analysis of a Multiaccess Control Scheme", *IEEE Transactions on Automatic Control*, Vol. AC-32, No. 11, November 1987, pp. 1017 – 1020.

[WU93]   Wu, T. and Chang, J., "Collision Resolution for Variable Length Messages", *IEEE Transactions on Communications*, Vol. 41, No. 9, September 1993, pp. 1281 – 1288.

Chapter 6

SPREAD SPECTRUM MULTIPLE ACCESS

The multiple access technique in spread spectrum networks usually refers to the ability of certain kinds of signals to coexist in the same frequency and time space with an acceptable level of mutual interference. The use of pseudorandom or pseudonoise (PN) waveforms in a wireless network is motivated largely by the desire to achieve good performance in fading multipath channels and the ability to operate multiple links with pseudo-orthogonal waveforms using spread spectrum multiple access. Typically, a RAKE receiver achieves multipath diversity by exploiting the large bandwidth inherent in spread spectrum systems. Unlike FDMA and TDMA, every user in a SSMA system is allocated the full bandwidth all the time.

6.1 SPREAD SPECTRUM COMMUNICATIONS

Spread spectrum has been used in many military applications including anti-jamming, ranging and secure communications. It refers to signaling schemes which are based on some form of coding (that is not a function of the transmitted information) and which use a bandwidth that is several orders of magnitude greater than the information rate. The bandwidth expansion is achieved using a pseudonoise (PN) code that converts a narrowband signal into a noise-like signal before transmission. The PN code is a special sequence of bits which are called chips because they represent small sections of a data bit. The sequence repeats itself after a finite amount of time. This sequence possesses desirable correlation properties (i.e., strong autocorrelation compared with itself and low cross-correlation when compared with other codes) that enable a spread spectrum receiver to recover the intended information signal even when other users are transmitting using the same bandwidth at the same time.

Spread spectrum is more resistant to multipath effects and more tolerant of interference. They are primarily interference (or power) limited rather than bandwidth limited. Since spread spectrum spreads the energy over a large bandwidth, the energy per unit frequency is correspondingly reduced by the same factor. Hence, the interference produced is significantly smaller as compared to narrowband systems. There is a fundamental difference between the bandwidth expansion due to coding and that due to spectrum spreading. Spectrum spreading plays no role in increasing channel capacity but can perform other useful roles such as providing low probability of interception of the signal, good electromagnetic capability, and a multiple access capability.

There are generally two types of SSMA techniques:

❑   frequency-hopped spread spectrum (FHSS);
❑   direct-sequence spread spectrum (DSSS).

Phase modulation is employed by DSSS systems whereas FHSS systems employ frequency modulation. A DSSS signal occupies all of the bandwidth at a fraction of the of the original signal power level while a FHSS signal occupies a fraction of the bandwidth for a fraction of the time at the original power level. As a result, the interference in DSSS systems is continuous which is in contrast to the pulsed-type interference emitted by FHSS systems.

In the US, regulations for these spread spectrum techniques exist. For example, FCC requires the PN code used in DSSS systems to be at least 10 chips in length. Regulations for FHSS systems depend on the operating frequency (Table 6.1).

Table 6.1: FCC regulations for FHSS systems

Specification	900 MHz	2.4 GHz
Maximum channel width	500 KHz	1 MHz
Maximum dwell time on a channel	400 ms (every 20 s)	400 ms (every 30s)
Minimum number of channels	50 (out of 52)	75 (out of 83)

6.2 WIDEBAND VERSUS NARROWBAND

The multiple access capability of wideband spread spectrum systems is distinctly different from narrowband systems. Not only can simultaneous transmissions be tolerated, the number of such transmissions should be large in order to achieve high network capacity. Thus, as long as different receivers are involved, SSMA systems are designed to have multiple transmissions taking place at the same time. In this case, interference may sometimes dominate over noise as an error-producing mechanism.

Among the many advantages, SSMA provides soft capacity where no absolute limit is placed on the number of transmitting users. Performance is degraded in proportion to any increase in the number of transmitting users [GILH91]. Thus, SSMA systems are interference limited. This is in contrast to narrowband systems, which are primarily bandwidth limited. The key features that differentiate SSMA systems from narrowband systems are code division, capture, and packet collisions.

6.2.1 Code Division

Code division refers to the fact that transmission with orthogonal PN codes may overlap in time with little or no effect on each other. Different users transmit using unique codes. For a receiver tuned to the code of one transmission, other signals appear as noise. This is possible because the interference sources are not correlated with the desired signal. In the dispreading (decoding) process, this noise will be suppressed by the processing gain.

6.2.2 Time Capture

Capture in SSMA systems refers to the ability of a receiver to successfully receive a packet with a given PN code despite the simultaneous presence of other time-overlapping signals with the same code. Usually, the packet captured by the receiver corresponds to either the first arriving signal (time capture) or the strongest signal (power capture) [TOBA84]. While both SSMA and narrowband systems are capable of power capture, the time capture property is only unique to SSMA systems.

Without power capture, any slight overlap of two or more packets in a narrowband system results in the destruction of all, regardless of which receiver the packet is intended for. This is because the received signal is designed to provide close to one-bit per Nyquist sample [ABRA94]. However, in spread spectrum systems, much less than one bit of information per sample is provided by the signal. Thus, the receiver has to combine a large number of Nyquist samples for each received bit. As a result, packets from two different transmitters may overlap in time without necessarily leading to complete loss (Figure 6.1).

6.2.3 Collisions in SSMA Systems

A collision in a SSMA system has about the same effect as a collision in a narrowband system. Packets involved in the collision are discarded and must be retransmitted at a later time. An important difference that arises in a SSMA system is the possibility that even when two packets collide, yet, at the same time, a third packet is received (e.g., the third packet may be using a different PN code). Thus, with SSMA, collision events are associated with individual packets rather than with time slots or with individual receivers. Moreover, the mechanism that leads to collisions should not be confused with the random errors that may occur when multiple packets are transmitted simultaneously, perhaps to different receivers, using different PN codes or time offsets of the same code [PURS87].

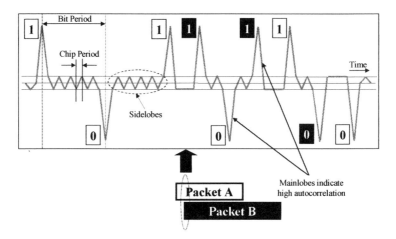

Figure 6.1: Overlapped transmissions in time capture

There are two types of multiuser interference that result in spread spectrum collisions. Primary collisions occur when two or more users transmit simultaneously using the same PN code. Secondary collisions refer to the interference that results from signals using codes that are quasi-orthogonal to that of the desired signal. Thus, a packet will be received correctly only if it does not suffer too many secondary conflicts.

In order for a spread spectrum signal to be successfully received, it is necessary for the receiver to know the PN code and to monitor it at the right time. Thus, there must be coordination between the receiver and transmitter. This situation suggests a third type of interference, which can be termed receiver scheduling conflicts. If two users transmit to a common destination simultaneously using different codes (assuming that the destination knows that these users are transmitting and also the codes that they are using), the receiver will be able to monitor only one of them even though there are no primary conflicts and the secondary interference may be sufficiently low.

6.3 DIRECT SEQUENCE SPREAD SPECTRUM

In DSSS systems, a PN code generator produces a sequence of binary chips which is n times faster than the data rate. Each data bit is modulo-two added with this sequence to form a DSSS signal, thus spreading the bandwidth by a factor of n (Figure 6.2). The receiver uses the same PN code to perform time-correlation detection (or match filtering). As the incoming signal is processed, whenever there is a match with the desired PN code, the autocorrelation function will exhibit a main lobe, narrow in width and higher in amplitude than that obtained when there is no match.

Simple systems employ fixed-length PN code patterns for all users. More sophisticated systems employ bit-by-bit changing PN codes in conjunction with programmable match filters at the receiver. Such systems enjoy enhanced time capture probability. Note that the difference between the total number of high and low voltage transitions in any PN code differs by at most one. This property helps to maintain the average transmitted voltage at close to zero.

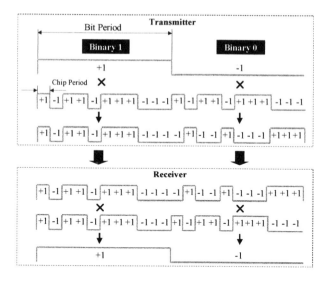

Figure 6.2: DSSS modulation

6.3.1 Processing Gain

The wide bandwidth in DSSS transmission provides the means to separate multipath signals using correlators or matched filters. Usually, only one of the multipath signals is synchronized. Signals that are delayed by more than one chip duration are significantly attenuated. Once separated, the signal components can be combined to reduce fading over time and improve the signal-to-noise ratio. The signal-to-noise ratio is improved by a factor called the processing gain (K) where

$$K = 10 \log n \quad \text{dB}$$

(6.1)

6.3.2 Direct Sequence Code Division Multiple Access

DSSS is usually not bandwidth efficient when used by a single user. This gives rise to direct sequence code division multiple access (DS-CDMA) where different PN codes are assigned to different users. Figures 6.3 to 6.6 illustrate how these codes allow multiple users to access the channel simultaneously without interfering.

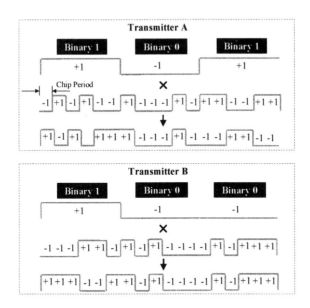

Figure 6.3: Employing different PN codes in DS-CDMA

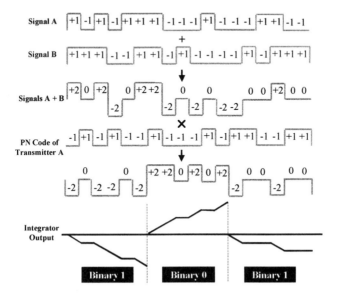

Figure 6.4: Decoding in DS-CDMA (Transmitter A)

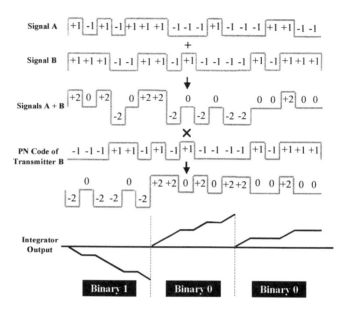

Figure 6.5: Decoding in DS-CDMA (Transmitter B)

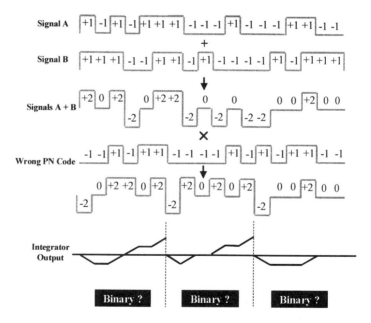

Figure 6.6: Erroneous decoding in DS-CDMA

6.3.3 Power Control

In narrowband systems, a difference in the received power has a positive effect on the throughput because when several packets collide, the one that is closest to the receiver may survive the collision through power capture. In a DS-CDMA environment with simultaneous transmissions, the situation is different. The transmitters that are closer to the receiver interfere strongly with those farther away and prevent them from communicating properly, thereby reducing the number of simultaneous users. This is known as the near/far problem. Note that TDMA systems do not suffer from this problem since only one transmitter is active at any one time.

Since the presence of other interfering signals prevent small signals from being demodulated properly, this means that a DS-CDMA system is limited by the weakest transmitter. For example, when a user nears the edge of a radio cell, every other user must reduce its transmitted power to accommodate this weak user and this in turn can cause every receiver to reach its threshold. Conversely, if many users are transmitting at peak power from the edge of the cell, the performance of the cell can be severely degraded. Thus, DS-CDMA systems require the power of all transmitting users to arrive at the receiver at about the same level. In this way, interference from all users in the network is kept to a minimum. For cellular systems, the objective is to keep the received signal power at the base station fixed at the desired level.

Two types of power control are normally used for uplink transmission to the base station. In open-loop power control, the sum of the transmit and receive powers are kept to a constant. This means that a reduction in signal level at the receiving antenna will result in an increase in signal power from the transmitter. Such control assumes that signal loss in the downlink is identical to the uplink. In close-loop power control, the base station monitors the power received from each mobile terminal and then commands the terminal to increase or decrease its power. Thus, transmit power is minimized, resulting in longer battery life. A combination of close-loop and open-loop power control may also be employed. For example, the open-loop algorithm may deal with long-term fading due to shadowing while the close-loop power control reacts to fast fading caused by multipath propagation.

6.3.4 Synchronization

With random access (asynchronous) transmissions on a shared wireless channel, a receiver cannot anticipate the sender of any particular packet or its exact time of arrival. The receiver must use the initial portion (preamble) of a packet to detect the arrival of the packet, and acquire bit and packet synchronization. However, a PN code has many phase shifts and any of the chips in the code can be selected as the first chip in the packet. One function of the timing mechanism, therefore, is to determine which phase is to be employed for each transmitted packet. The design of the packet preamble and the corresponding receiver system suitable for achieving rapid synchronization to a received packet is a function of the particular PN waveform being used [PURS87].

Besides packet synchronization, base stations in a DS-CDMA cellular network need to distinguish themselves from each other by transmitting different portions of a common PN code at a given time. In other words, the base stations transmit time offset versions of the same PN code. In order to guarantee uniqueness of the time offsets, user terminals must remain synchronized to a common time reference. This can be achieved using the Global Positioning System (GPS), which provides a precise time reference. GPS is a satellite-based, radio navigation system capable of providing continuous position, velocity, and time information to an unlimited number of users. Through GPS, users can acquire and maintain a common network time of day. In doing so, receivers are able to obtain properly referenced PN waveforms even as a PN code varies with time.

6.3.5 RAKE Receiver

DS-CDMA takes advantage of multipath propagation by assigning multiple correlators (or adaptive match filters) to resolve and combine signals of one user received over different propagation paths. This multiple correlator system is called a RAKE receiver (Figure 6.7). The received signal is fed to a system of taps with each tap having a different delay. Any signal that matches the delay in a tap is passed to a weighting function that scales the signal in order to correct amplitude errors. After scaling, each tap is summed together to create a stronger version that matches the original signal more closely. The RAKE process is effective when multipath delays are longer than one chip duration.

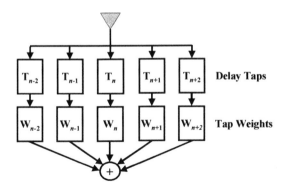

Figure 6.7: RAKE receiver design

6.3.6 Benefits of DSSS

Since the signal is spread over a wider bandwidth, the spectral density of the signal is reduced for a constant transmitted power level. This factor coupled with the PN nature of the waveform gives the system a lower electromagnetic profile and hence, the inherent privacy.

Spatial diversity in DSSS systems takes two forms. In dual antenna systems, the base station uses two receive antennas placed sufficiently far apart to mitigate fades. The more popular alternative is to allow multiple base stations communicate with the mobile terminal in a soft handoff. In this case, signals from multiple base stations overlap each other, making it possible for a user terminal to contact or more base stations simultaneously. This reduces the probability of an unsuccessful handoff. The signals from the base stations are treated as multipath signals and are coherently combined at the mobile terminal. Soft handoff enables "make-before-break" handoff, which is in contrast to the "break-before-make" handoff characteristic in TDMA systems.

DSSS systems enjoy frequency diversity since frequency-selective fading typically affect only a small portion of the signal spectrum, thus resulting only in a partial power loss in the received signal as opposed to a complete loss in narrowband signals. In addition, since the received energy is relatively constant over time, the overall communication reliability is similar to that of a frequency diversity system.

A third type of DSSS diversity involves time diversity where signals are spread in time using interleaving. Forward error correction is typically applied, along with maximal likelihood detection. The IS-95 standard for example, makes use of convolutional codes (1/2 rate on forward channel, 1/3 rate on reverse channel) with interleaving.

Finally, DSSS can take advantage of silent periods in a voice conservation to increase capacity. To achieve this, a variable rate vocoder is required.

6.4 FREQUENCY HOPPED SPREAD SPECTRUM

In FHSS, the total bandwidth is divided into subchannels called frequency slots with one carrier frequency available in each of slot. The bandwidth of any slot is much smaller compared to the total spread bandwidth. The slots that represent the information signal are randomly varied according to a predefined frequency-hopping pattern, which varies according to a PN code. Thus, the hopping pattern is independent of the information bits. Data symbols are modulated and transmitted on the sequence of slots. Clearly, symbols transmitted on the same frequency at the same time from two or more different users are destroyed. Thus, to be useful, the FHSS requires forward error correction (FEC) which allows the transmitted data to be recovered despite the occasional loss of data symbols. Note that TDMA-based GSM also uses frequency hopping techniques to provide lower power mobile terminals some diversity gain at the edge of radio cells.

6.4.1 Slow and Fast FH

The ratio between the hopping rate and the data rate results in two modes of FHSS systems (Figure 6.8). When the hopping rate is higher than the data rate, then the system is known as fast frequency hopping. Conversely, when the hopping rate is lower (as in the example above), then the system is classified as slow frequency hopping.

The hopping rate has a profound effect on the performance of a FHSS system. Unlike narrowband systems that operate in dedicated spectrum and are generally not concerned with interference, frequency hopping systems may experience interference on some channels since the PN codes that control the hopping patterns are not completely orthogonal to each other.

For slow FHSS systems, this can potentially lead to the loss of many data packets. Thus, fast frequency hopping usually outperform slow hopping, even with the same processing gain. However, fast FHSS are expensive to implement since they require very fast frequency synthesizers.

6.4.2 Benefits FHSS

The degree of interference rejection afforded by a FHSS depends on the hopping rate and on the type of error correction employed. The basic concept is that data may be corrupted on some frequencies, but this is compensated by having most of the data correctly received on other frequencies and by allowing FEC restore the lost data. Suppose interference is distributed uniformly over all n hopping frequencies. Then FHSS may realize an average signal-to-noise ratio improvement given by equation 6.1.

Although changing in frequency, FHSS systems remain narrowband with excellent signal-to-noise performance and interference rejection afforded by narrow filters. It is also not prone to the near/far problem that plaques DSSS systems. Moreover, synchronization is performed at the data rate rather than the much faster chipping rate as in DSSS systems.

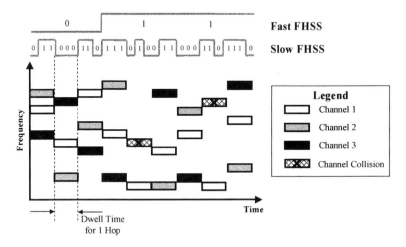

Figure 6.8: Fast and slow FHSS

6.5 SPREADING CODE PROTOCOLS

In spread spectrum systems, each user cannot arbitrarily choose a PN code for a given transmission without network considerations because two users employing the same PN code at the same time will experience a collision. Predefined protocols called spreading code protocols are therefore required. Users specify which codes to monitor when idle and which code to use when transmitting data. Spreading code protocols may be classified under common code, receiver-based, transmitter-based, or hybrids of these [PURS87].

6.5.1 Common Code

The use of a single common code is essentially equivalent to operating a single channel because a transmission is successful only if no other user transmits at the same time. Such a code assignment clearly fails to exploit the multiuser capability offered by SSMA.

6.5.2 Transmitter-Directed Codes

In this case, every user is assigned a unique transmitter-based code. When a user transmits, the intended receiver must monitor the code used by the transmitter. Clearly, transmitter-directed codes place a considerable burden on the receiver. They are ideal for broadcast transmission because many users can monitor the same code. Such codes also allow selective reception even when several active transmissions are taking place at the same time [EPHR87]. An important property of transmitter-based codes is that their use precludes primary conflicts since no two users will ever use the same code.

6.5.3 Receiver-Directed Codes

A receiver-directed code can be thought of as an address that identifies the receiver. A user always monitors its own code regardless of the transmitter's identity. The transmitter, however, must use the code of the intended receiver. Information regarding the assignment of codes to receivers must be disseminated throughout the network.

6.5.4 Single Code for Pairs of Users

In this case, every pair of users is assigned a unique code. Thus, if user A wishes to transmit to user B, user A must use the specific code assigned to that pair and user B must monitor the same code at the same time. This scheme is clearly inappropriate for broadcast traffic.

6.6 CDMA NETWORK DESIGN

From an operations standpoint, CDMA was designed to be simple to optimize because it uses a frequency reuse of one (i.e., the same frequency is used in every radio cell). Unlike TDMA systems, which use conventional frequency reuse techniques for cell planning, CDMA's reuse frequency of one makes it easier to plan the network. However, practical implementations of CDMA networks have been found to be difficult and time-consuming to optimize, particularly during heavy traffic loads due to the extra interference generated by the increased number of users.

Effective CDMA network design hinges on three primary factors namely, coverage, quality, and capacity. Since these factors are tightly inter-related, they must be balanced off from each other to arrive at the desired level of system performance. Thus, higher capacity can be achieved through some degree of degradation in coverage and/or quality.

6.6.1 Multirate CDMA

CDMA systems are capable of supporting a wide range of bit rates. Due to the requirements imposed by changing network conditions, it is desirable to provide a means for trading data rate for processing gain. For example, in an environment with severe interference or fading, the communication link quality can be improved by employing a lower data rate to achieve the necessary increase in processing gain. This implies that variable spreading factors and multiple chip rates must be incorporated into the system.

The service bit rates can be varied either by varying the spreading factor while maintaining a constant spread bandwidth or by varying the number of codes transmitted in parallel, each with a fixed spreading factor. Alternatively, multiple PN codes can be assigned to the user. A new version of IS-95B for example, assigns up to eight codes. Such code aggregation allows high-speed data transmission. A similar approach involves multirate code division where each user accesses the link using a spreading factor equal to the available bandwidth divided by the required peak data rate.

6.6.2 CDMA Receivers

In a wireless channel, both intersymbol interference (ISI) and multiple access interference (MAI) occur at the receiver. This gives rise to two types of CDMA receivers: one that treats ISI and MAI as noise and the other, exploiting some or all knowledge about ISI and MAI. The latter method is made possible by predefined PN codes (which provide information about MAI) and channel estimates (which provide information about ISI).

6.6.3 Multiuser Detection

In multiuser detection, the signals from all users are jointly detected by exploiting knowledge about the predefined PN codes as well as the channel impulse response. Thus, intracell interference is virtually eliminated and the resulting interference situation is similar to that of FDMA and TDMA systems. Note that multiuser detection allows a receiver to capture more than one packet that are overlapped in time. This capability may give future CDMA systems the leading edge over TDMA systems. However, the simultaneous detection of multiple signals requires low bit error rates because bits that are erroneously detected are subtracted from the signals of other users, potentially causing those signals to be decoded in error as well.

Multiuser detection removes the need for sophisticated, high-precision power control in CDMA systems by exploiting the deterministic structure of MAI. This is possible since the interference in spread spectrum environments is a significantly less random than Gaussian noise. An equally important advantage is the capacity improvements even in situations of exact power control (i.e., equal power reception). This performance gain results in lower power consumption and processing gain, which translate into increased battery life and lower bandwidth requirements.

The optimum multiuser detector is unfortunately too complex to be implemented. Suboptimum detectors are therefore proposed. An important class of such suboptimal detectors with linear complexity (with respect to number of users) is the decorrelating detector. The decorrelating detector can be compared to the zero-forcing receiver (for single user transmission over ISI channels) but in this case, it completely eliminates all synchronous MAI. Interference from the same user as well as MAI from other users are removed via the use of decision feedback.

6.6.4 Interference Cancellation and Suppression

Interference cancellation refers to the subtraction of interference either by serial (successive) or parallel cancellation. Interference cancellation relies on first detecting the strongest signal in the composite received signal and then reconstructing the contribution of this signal before subtracting it from the composite signal. In doing so, the effective interference on the other residual signals in the composite received signal is reduced. The procedure is then repeated for the second strongest signal and so on. Unlike parallel cancellation, successive cancellation simplifies the design of multiuser detectors. Such approaches are able to successively cancel out the strongest interfering users.

Interference suppression using adaptive signal processing may have an advantage over interference cancellation schemes since it eliminates the need for accurate estimation of the received signal's amplitude and phase.

In both methods, acquisition considerations (e.g., employing optimal and suboptimal receiver structures) and the use of adaptive spatial antenna arrays play central roles.

6.6.5 Multicarrier CDMA

A number of wideband and hybrid CDMA schemes [MORI97] have been proposed for wireless broadband communications. These schemes are depicted in Figure 6.9. As explained in Section 1.7, multicarrier schemes employ parallel signaling methods that offer several advantages over conventional single carrier systems such as protection against dispersive multipath links and frequency-selective fading. With appropriate signal processing and forward error correction coding, these systems can achieve the equivalent capacity and delay performance of single carrier systems without the need for a continuous frequency band. Multicarrier schemes, when used in conjunction with OFDM, are bandwidth-efficient since guard bands between adjacent carriers are unnecessary in OFDM.

6.7 PERFORMANCE ANALYSIS

CDMA networks have been analyzed extensively. The analysis is mostly centered on increasing the throughput of the system by

❑   reducing the detection threshold;
❑   employing convolutional coding on the information bit stream;
❑   employing block coding over the entire packet.

Other analysis has dealt with the mutual interference of the users resulting from the use of non-perfect codes. There is however, no one set of analytic tools that can simultaneously account for the communication theory and communication network aspects of a CDMA system. Analysis which address the communication theory aspects of the system (e.g., probability of bit error within a packet) model the network as a single Poisson stream of packets with the users being grouped as transmitter-receiver pairs but cannot account for realistic traffic models. On the other hand, network-based analysis takes transmit/receive conflicts and the random pairing of users as transmitter-receiver pairs into account but cannot model the system behavior at the bit level. Unfortunately, both types of analysis are necessary in the study of spread spectrum networks. However, any model which attempts to account for the two system aspects simultaneously quickly becomes intractable [SOUS88].

6.7.1 Queuing Theory Modeling

Generally, the implementation of narrowband access protocols results in the use of the entire channel bandwidth for a given transmission. Thus, in analytical work, there are only two components to concentrate on:

❑   a channel, through which all the successful packets pass (one at a time);
❑   a collection of users, which may be lumped together to form an arrival process.

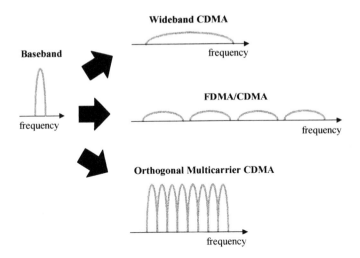

Figure 6.9: Spread spectrum multiple access techniques

The throughput analysis reduces to performing an operation on a predefined input random process and obtaining some statistics of the output process.

With spread spectrum signaling, however, the possibility of simultaneous successful transmissions precludes the above characterization since there is no single channel through which all packets pass through. Instead of the above characterization, which resembles a single server queue, a spread spectrum characterization resembles a complex interaction among a set of queues. Thus, the performance analysis of a spread spectrum network is an extremely difficult task due to the fact that the system is driven by a set of uncoordinated users that interfere with each other to various degrees. Furthermore, the notion of collisions have to be redefined since transmitters can use the same PN code and the same interval provided their relative delays are greater than one chip time.

6.7.2 Estimating the Number of Users in DS-CDMA

To determine the number of users for an asynchronous DS-CDMA system using BPSK modulation, one may treat interference from all interfering users as independent, additive white Gaussian noise. This is a good approximation provided:

❑   the PN code is long;
❑   the number of users is large.

For many CDMA systems where the signals are not independent, the Gaussian approximation is also justified although the variance may be different [PAHL85].

Suppose there are N users operating with a processing gain of K. The received power (S) from each user is assumed to be the same. Then, the signal-to-noise ratio before the correlator is

$$\frac{S}{(N-1)S} = \frac{1}{N-1}$$

(6.2)

After correlation, the peak power of the desired signal is increased by a factor of K while the statistics of the noise (interference) created by the other interfering users remain the same. The signal-to-noise ratio after the correlator is therefore given by the bit energy-to-noise ratio (E_b/N_0). E_b is obtained by dividing the desired signal power by the information rate (R). N_0 represents the total noise contribution in the spreading bandwidth (W). If background noise due to spurious interference and thermal noise is ignored, then E_b/N_0 can be approximated by

$$\frac{E_b}{N_o} = \frac{S/R}{[(N-1)S]/W} = \frac{W/R}{N-1} = \frac{K}{N-1}$$

(6.3)

Equation 6.2 provides the worst case E_b/N_0 where all interfering signals are aligned in time. Usually, if the transmitted signals are not aligned the amount of interference reduces by a factor of three, yielding:

$$\frac{E_b}{N_o} = \frac{3K}{N-1}$$

(6.4)

$$\therefore N = \frac{3K}{E_b/N_o} + 1$$

(6.5)

For BPSK modulation, the bandwidth efficiency is N/K. This number can be compared with the number of users supported with standard multiple access techniques (e.g., TDMA, FDMA) for the same allocated bandwidth.

6.8 COMPARING FDMA, TDMA, AND CDMA

Under an environment where single path channels are corrupted only by Gaussian noise, the channel capacity of FDMA, TDMA, CDMA, and hybrids of these schemes are equivalent [BAIE96]. Differences arise when additional channel characteristics such as multipath propagation and frequency selective fading are factored in. The major differences are summarized in Table 6.2 [NIEL85].

6.8.1 Advantages of CDMA

The main problem with TDMA and FDMA is that bandwidth resources are left idle even when there are packets waiting to be transmitted. This problem is particularly acute for users with bursty traffic. In this case, CDMA is attractive for the following reasons:

❑   By foregoing the rigid time scheduling requirement in TDMA, the possibility of concurrent transmissions is achieved;
❑   By not being allocated a fixed frequency band as in FDMA, CDMA radio transceivers can switch flexibly from one signal to another.

In general, a properly designed CDMA system can provide more capacity than an equivalent TDMA system [GILH91]. This capacity increase may be attributed to the powerful spatial reuse effect in CDMA which allows reuse of the same bandwidth in all radio cells, thereby increasing capacity in terms of the number of users/unit area/unit bandwidth. Such spatial reuse more than offsets the initial bandwidth penalty due to spreading. However, the efficiency of spatial reuse is dependent on a number of factors including the propagation loss, the multiplexing efficiency of the PN code, and the modulation technique employed.

The capacity of FDMA and TDMA is strictly defined by the number of available frequency channels and time slots. In contrast, the capacity of CDMA systems is a flexible parameter determined by the amount of interference and required service quality. Statistical multiplexing is inherent in such systems since reduced interference (when users are idle) allows other users to access the system or operate at a higher data rate. As the number of transmitting users increases, the signal-to-noise ratio becomes smaller, and the performance of each receiver declines in a gradual manner. Thus, CDMA provides better adaptability to changing network traffic loads.

Table 6.2: Comparison of FDMA, TDMA, and CDMA

Condition	FDMA	TDMA	CDMA
Requirement for maximum spectral efficiency	Zero adjacent channel interference (requires non-realizeable brick-wall filter)	Zero timing error (impossible for distributed users)	Orthogonal codes for all time shifts (cannot be found)
Practical requirement for good spectral efficiency	Frequency guard band	Time guard band	Pseudo-orthogonal codes (results in added noise)
Transmission characteristics	Continuous	Intermittent (require buffers)	Continuous
User data rate	Users with different data rates are assigned different bandwidth	Users with high data rates are assigned more time slots	Most efficient for users with one data rate or a multiple of it
System requirement	Frequency band assignment	Time slot assignment, network synchronization	PN code assignment
Flexibility of adding new users	No	No	Yes, with graceful degradation
Transmit power	Average power	Peak power	Average power
Multipath resistance	Vulnerable	Vulnerable	Inherent
Broadcast capability	A common frequency band is required	Inherent	A common PN code is required

Another benefit of CDMA is that it can potentially support overlay operations over existing radio networks when new frequencies are not available. Furthermore, interference between CDMA users is always noise-like. Such interference is more acceptable than the intermodulation interference that may occur in TDMA and FDMA systems.

CDMA systems, being wideband systems, provide greater immunity to multipath fading. In CDMA, multipaths can actually be utilized to enhance the detection process. The time-varying characteristics of the multipath channel can be estimated in a better way in CDMA receiver than in FDMA or TDMA since the wider bandwidth provides more information. For TDMA, the multipath propagation of the signal from a certain slot transmission can affect neighboring slots, resulting in data errors unless expensive equalizers are built and trained. Alternatively, large gaps between slots can reduce multpath interference but this leads to wasted bandwidth resource.

A unique advantage of CDMA is the soft handoff capability. When CDMA connections need to be moved from one radio cell to another, each connection is maintained by base stations at two or more nearby cell cites. Thus, unlike TDMA systems which use hard handoff techniques, CDMA is less likely to drop connections. Besides increased handoff reliability, handing off a connection to a congested CDMA cell does not necessarily result in connection blocking or termination. Instead, switching to a highly loaded cell typically leads to a higher interference level but the communications link is maintained.

Perhaps the most compelling advantage of CDMA schemes compared to TDMA is that there is no precise time coordination. CDMA does not require distinctly defined channels. All users employ the same channels and no rigid time structure is required in accessing the channel. This removes the need for guard times and the need to synchronize all transmitting users. Synchronization is a serious problem for TDMA when mobile transmitters are distributed across different geographical areas.

6.8.2 Disadvantages of CDMA

The advantages of CDMA are offset by several disadvantages. Specifically, DS-CDMA systems require accurate power control to avoid the near/far problem. Power control ensures that the signal from each user arrives at the receiving antenna with approximately equal strength. Failure to control the power output in DS-CDMA may result in one user swamping every other user. This is in contrast to TDMA systems which can operate with loose power control since only one user transmits at any one time.

In FDMA and TDMA cellular networks, only intercell interference is present. In CDMA networks, both intercell and intracell interference sources have to be dealt with.

SUMMARY

In spread spectrum communications, any reduction in interference converts directly into an increase in network capacity. Thus, the exact probability of success of a packet in the presence of other signals is very difficult to evaluate. Several methods that have the potential of raising the capacity of spread spectrum systems significantly include multiuser detection, interference cancellation, and interference suppression.

BIBLIOGRAPHY

[ABRA93]   Abramson, N., *Multiple Access Communications*, IEEE Press, 1993.
[ABRA94]   Abramson, N., "Multiple Access in Wireless Digital Networks", *Proceedings of the IEEE*, Vol. 82, No. 9, September 1994, pp. 1360 – 1370.
[ARRL91]   American Radio Relay League, *The ARRL Spread Spectrum Sourcebook*, ARRL, 1991.
[BAIE93]   Baier, A. and Panzer, H., "Multirate DS-CDMA Radio Interface for Third Generation Cellular Systems", *Mobile and Personal Communications*, December 1993.
[BAIE94]   Baier, A., Fiebig, U., Granzow, W., Koch, W., Teder, P. and Thielecke, J., "Design Study for a CDMA-Based Third-Generation Mobile Radio System", *IEEE Journal on Selected Areas in Communications*, Vol. 12, No. 4, May 1994, pp. 733 – 741.
[BAIE96]   Baier, P., "A Critical Review of CDMA", *Proceedings of the 44th IEEE Vehicular Technology Conference*, 1994, pp. 1 – 5.

[CHEN91] Chen, X. and Oksman, J, "The Receiver and Transmitter Code Sensing Protocol and its Applications in Distributed CDMA Networks", *Proceedings of the IEEE INFOCOM*, 1991, pp. 1326 – 1333.

[CHOU85] Chou, W., *Computer Communications Volume 2: System and Applications*, Prentice Hall, 1985.

[COOK83] Cook, C. (editor), *Spread Spectrum Communications*, IEEE Press, 1983.

[DIXO94] Dixon, R, *Spread Spectrum Systems*, John Wiley, 1994.

[ELHA83] Elhakeem, A., Hafez, H., and Mahmoud, S., "Spread-Spectrum Access to Mixed Voice-Data Local Area Networks", *IEEE Journal on Selected Areas in Communications*, Vol. SAC-1, No. 6, December 1983, pp. 1054 – 1060.

[EPHR87] Ephremides, A., Wieselthier, J. and Baker, D., "A Design Concept for Reliable Mobile Radio Networks with Frequency Hopping Signaling", *Proceedings of the IEEE*, Vol. 75, No. 1, January 1987, pp. 56 – 73.

[GILH91] Gilhousen, K., Jacobs, I., Padovani, R., Viterbi, A., Weaver, L. and Wheatley, C., "On the Capacity of a Cellular CDMA System", *IEEE Transactions on Vehicular Technology*, May 1991, appearing in [ABRA93].

[GARG98] Garg, V., *Applications of CDMA in Wireless Personal Communications*, Prentice Hall, 1998.

[GLIS95] Glisic, S., (editor), *Code Division Multiple Access Communications*, Kluwer Academic Publishers, 1995.

[GLIS97] Glisic, S. and Leppänen, P., (editors), *Wireless Communications: TDMA versus CDMA*, Kluwer Academic Publishers, 1997.

[HUI83] Hui, J., *Fundamental Issues of Multiple Access*, PhD Dissertation, MIT, 1983.

[HUI84] Hui, J., "Throughput Analysis for Code Division Multiple Accessing of the Spread Spectrum Channel", *IEEE Journal of Selected Areas in Communications*, July 1984, appearing in [ABRA93].

[IEEE77] "Special Issue on Spread Spectrum Communications", *IEEE Transactions on Communications*, Vol. COM-25, No. 8, August 1977, pp. 745 – 747.

[IEEE87] "Special Issue on Packet Radio Networks", *Proceedings of IEEE*, Vol. 75, No. 1, January 1987.

[IEEE94a] "Special Issue on CDMA: CDMA I", *IEEE Journal on Selected Areas in Communications*, May 1994.

[IEEE94b] "Special Issue on CDMA: CDMA II", *IEEE Journal on Selected Areas in Communications*, June 1994.

[KAHN78] Kahn, R., Gronemeyer, S., Burchfiel, J. and Kunzelman, R., "Advances in Packet Radio Technology", *Proceedings of the IEEE*, Vol. 66, November 1978, pp. 1468 – 1496.

[KAVE87] Kavehrad, M., "Direct-Sequence Spread Spectrum with DPSK Modulation and Diversity for Indoor Wireless Communications", *IEEE Transactions on Communications*, Vol. COM-35, No. 2, February 1987, pp. 224 – 236.

[LEHN87] Lehnert, J. and Pursley, M., "Multipath Diversity Reception of Spread-Spectrum Multiple Access Communications", *IEEE Transactions on Communications*, Vol. COM-35, November 1987, pp. 1189 – 1198.

[LEIN87] Leiner, B., Nielson, D. and Tobagi, F., "Issues in Packet Radio Network Design", *Proceedings of IEEE*, Vol. 75, No. 1, January 1987, pp. 6 – 20.

[LIU96] Liu, Z., Karol, M., Zarki, M. and Eng, K., "Channel Access and Interference Issues in Multi-code DS-CDMA Wireless Packet (ATM) Networks", *ACM/Baltzer Journal on Wireless Networks*, Vol. 2, No. 3, 1996, pp. 173 – 193.

[MORI97] Morinaga, N., Nakagawa, M. and Kohno, R., "New Concepts and Technologies for Achieving Highly Reliable and High Capacity Multimedia Wireless Communications Systems", *IEEE Communications Magazine*, Vol. 35, No. 1, January 1997, pp. 34 – 40.

[MUSS86] Musser, J. and Daigle, J., "Derivation of Asynchronous Code Division Multiple Access Throughput", appearing in [PICK86].

[NIEL85] Nielson, D., "Packet Radio: An Area-Coverage Digital Radio Network", appearing in [CHOU85].

[PAHL85] Pahlavan, K., "Wireless Communications for Office Information Networks", *IEEE Communications Magazine*, Vol. 23, No. 6, June 1985, pp. 19 – 27.

[PICK82] Pickholtz, R., Schilling, D. and Milstein, M., "Theory of Spread Spectrum Communications – A Tutorial", *IEEE Transactions on Communications*, Vol. COM-30, No. 5, May 1982, pp. 855 – 885, also appearing in [ABRA93].

[PICK86] Pickholtz, R., *Local Area and Multiple Access Networks*, Computer Science Press, 1986.

[POLY87]   Polydoros, A. and Silvester, J., "Slotted Random Access Spread Spectrum Networks : An Analytical Framework", *IEEE Journal of Selected Areas in Communications*, July 1987, appearing in [ABRA93].

[PURS77]   Pursley, M., "Performance Evaluation for Phase-Coded Spread Spectrum Multiple Access Communication – Part I: System Analysis", *IEEE Transactions on Communications*, Vol. COM-25, August 1977, pp. 795 – 799.

[PURS87]   Pursley, M., "The Role of Spread Spectrum in Packet Radio Networks", *Proceedings of the IEEE*, January 1987, appearing in [ABRA93].

[RAYC81]   Raychaudhuri, D., "Performance Analysis of Random Access Packet-Switched Code Division Multiple Access Systems", *IEEE Transactions on Communications*, Vol. COM-29, No. 6, June 1981, pp. 895 – 901.

[SCHI91]   Schilling, D., Milstein, L., Pickholtz, R., Bruno, F., Kanterakis, E., Kullback, M., Erceg, V., Biederman, W., Fishman, D. and Salerno, D., "Broadband CDMA for Personal Communications Systems", *IEEE Communications Magazine*, November 1991, pp. 86 – 93.

[SCHO77]   Scholtz, R., "The Spread Spectrum Concept", *IEEE Transactions on Communications*, Vol. COM-25, No. 8, August 1977, pp. 748 – 755.

[SIMO94]   Simon, M., Omura, J., Scholtz, R. and Levitt, B., *Spread Spectrum Communications Handbook*, McGraw Hill, 1994.

[SOBR93]   Sobrinho, J. and Brazio, J., "C-MCMA : A New Multiple Access Protocol for Centralized Wireless Local Networks", *Proceedings of the ICC*, 1993.

[SOUS88]   Sousa, E. and Silvester, J., "Spreading Code Protocols for Distributed Spread Spectrum Packet Radio Networks", *IEEE Transactions on Communications*, Vol. 36, No. 3, March 1988, pp. 272 – 281.

[SOUS92]   Sousa, E., "Performance of a Spread Spectrum Packet Radio Network Link in a Poisson Field of Interferers", *IEEE Transactions on Information Theory*, Vol. 38, No. 6, November 1992, pp. 1743 – 1754.

[STOR89]   Storey, J. and Tobagi, F., "Throughput Performance of an Unslotted Direct-Sequence SSMA Packet Radio Network", *IEEE Transactions on Communications*, Vol. 37, No. 8, August 1989, pp. 814 – 820.

[TANT98]   Tantaratana, S. and Ahmed, K., *Wireless Applications of Spread Spectrum Systems: Selected Readings*, IEEE Press, 1998.

[TOBA84]   Tobagi, F., Binder, R. and Leiner, B., "Packet Radio and Satellite Networks", *IEEE Communications Magazine*, November 1984, appearing in [ABRA93].

[TURI80]   Turin, G., "Introduction to Spread-Spectrum Antimultipath Techniques and Their Application to Urban Digital Radio", *Proceedings of the IEEE*, Vol. 68, No. 3, March 1980, pp. 328 - 353.

[VERD86]   Verdu, S., "Minimum Probability of Error for Asynchronous Gaussian Multiple Access Channels", *IEEE Transactions on Information Theory*, Vol. IT-32, January 1986, pp. 85 – 96.

[VERD98]   Verdu, S., *Multiuser Detection*, Oxford University Press, 1998

[VERD00]   Verdu, S., "Wireless Bandwidth in the Making", *IEEE Communications Magazine*, July 2000.

[VITE79]   Viterbi, A., "Spread Spectrum Communications – Myths and Reality", *IEEE Communications Magazine*, Vol. 17, No. 3, May 1979, pp. 11 – 18.

[VITE85]   Viterbi, A., "When Not to Spread Spectrum – A Sequel", *IEEE Communications Magazine*, Vol. 23, No. 4, May 1979, pp. 12 – 17.

[VITE95]   Viterbi, A., *CDMA*, Addison Wesley, 1995.

[WIES83]   Wieselthier, J. and Ephremides, A., "A Distributed Reservation Scheme for Spread Spectrum Multiple Access Channels", *Proceedings of IEEE GLOBECOM*, 1983, pp. 659 – 665.

[WILS94]   Wilson, N., Ganesh, R., Joseph, K. and Raychaudhuri, D., "Packet CDMA Versus Dynamic TDMA for Multiple Access in an Integrated Voice/Data PCN", *IEEE Journal on Selected Areas in Communications*, August 1993, pp. 870 – 883.

[YOON93]   Yoon, Y., Kohno, R. and Imai, H., "A Spread-Spectrum Multiaccess System with Cochannel Interference Cancellation for Multipath Fading Channels", *IEEE Journal on Selected Areas in Communications*, Vol. 11, No. 7, September 1993, pp. 1067 – 1075.

[ZHAN93]   Zhang, Z. and Liu, Y., "Performance Analysis of Multiple Access Protocols for CDMA Cellular and Personal Communications Services", *Proceedings of the IEEE INFOCOM*, 1993, pp. 1214 – 1221.

Chapter 7

RESERVATION PROTOCOLS

Most contention protocols are suitable for low-rate bursty users with short messages. In contrast, reservation protocols are ideal for users that generate long, variable-length messages or users with slowly varying traffic. Reservation protocols usually require the broadcast channel to be divided into a reservation channel and a message channel. Each time slot in the message channel can accommodate one or more data packets and minislots are usually interleaved with the data slots so that short reservation requests can be transmitted. Higher throughputs are therefore possible in reservation access schemes because less bandwidth is wasted when requests collide in the minislotted channel and the maximum throughput of the message channel is unity. Reservation protocols trade higher utilization for higher connection setup latency.

7.1 GENERAL CHARACTERISTICS

Reservation protocols differ primarily in the way reservations are made and released. While there is no conflict in the use of the data channel, the multiple access problem is now shifted to the reservation channel. The difference is that a reservation request is short compared to the message and hence, the overhead required for conflict resolution is reduced. Once a reservation is successful, the user will usually not compete with other users with new or unsuccessful reservations and this in turn, enhances the chance of a successful reservation for these users.

Like TDMA, a common feature in most of these schemes is the division of time into slots which are then grouped into frames. Ready users request reservation for future time slots. The reservation channel can be based on fixed assignment or random access. The duration of each frame is greater than the propagation time. This enables each user to be aware of the usage of time slots in the previous frame. This is particularly important for satellite systems where propagation delays are long. Another underlying property of reservation protocols is the creation of a single global queue into which all packets are statistically multiplexed and serviced by the entire channel bandwidth. The minimum delay incurred by a packet, excluding packet transmission time, is at least three times the propagation delay. This can be significant when each packet is short.

7.2 CENTRALIZED VERSUS DISTRIBUTED

The presence of a centralized controller eliminates the queue synchronization problem. The central controller manages the global queue, accepts reservations, and informs users when to access the channel. In addition, priorities with efficient scheduling mechanisms can be implemented. If a central controller is employed in the reservation system, an additional subchannel for controller-to-user traffic is typically required.

For a distributed control approach, the queue status of all users must be synchronized. This means that reservation requests in the reservation channel need to be received correctly by all users. In the event of an error, the user must be able to reacquire queue synchronization (within a reasonable duration of time) by observing the exchange of messages in the reservation and message channels.

7.3 RESERVATION-ALOHA

R-ALOHA combines TDMA with slotted ALOHA. Channel transmission time is organized into frames comprising equal-length data slots which are identified by their positions in the frame [CROW73]. The number of slots per frame is generally less than the total number of users. These slots are either available or reserved. Reservations are implicit in that a successful transmission in any data slot serves as a reservation for the same slot in the next frame (Figure 7.1). By repeated use of the slot position, a user can transmit a long stream of data. This is equivalent to assigning the user a fixed TDMA channel. Any slot that is idle or contains a collision is available for contention using slotted ALOHA in the next frame. Thus, users need to maintain a history of the usage of each slot for one frame duration.

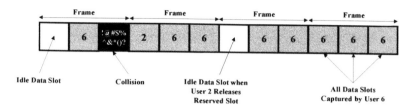

Figure 7.1: Operation of R-ALOHA

Clearly, the R-ALOHA protocol adapts itself to the nature of traffic. On one extreme, it behaves like slotted ALOHA when the users generate bursty traffic. At the other extreme, it behaves like TDMA when users generate continuous traffic. However, if a significant traffic imbalance exists (even for a short period of time), it is possible for a single user to capture all the available slots. This problem can be solved by requiring that interleaving empty slots be used.

7.3.1 Implementation Considerations

When a user gives up its reserved slot, a time slot is wasted in the following frame. To solve this problem, an end-of-use indicator can be included in the header of the last packet before a user gives up the slot. Priority can be incorporated by having users use an empty slot with a probability p. Higher priority users are given higher values of p.

7.3.2 Performance Considerations

Suppose v represents the average number of packets that a user transmits before giving up a reserved slot. Under the assumption of equilibrium conditions, the channel throughput (S_{RA}) of R-ALOHA can be expressed in terms of the slotted ALOHA throughput (S_{SA}) [LAM83], [CRIS95]:

Without end-of-use indicator:

$$S_{RA} = \frac{S_{SA}}{S_{SA} + 1/v}$$

$$(7.1)$$

With end-of-use indicator:

$$S_{RA} = \frac{S_{SA}}{S_{SA} + (1 - S_{SA})/v}$$

$$(7.2)$$

Since S_{RA} is a monotonically increasing function of S_{SA} and v ranges from one to infinity, the maximum throughput for R-ALOHA can be obtained by substituting $S_{SA} = 1/e$ and $v = 1$, thus giving:

Without end-of-use indicator:

$$\frac{1}{1+e}(\text{or } 0.269) \leq S_{RA,max} \leq 1$$

$$(7.3)$$

With end-of-use indicator:

$$\frac{1}{e}(\text{or } 0.368) \leq S_{RA,max} \leq 1$$

$$(7.4)$$

Note that without the use of the end-of-use indicator, the maximum throughput of R-ALOHA can be less than that of slotted ALOHA. This occurs when users have single-packet messages for transmission and this results in wasted time slots when they give up their reserved slots.

7.4 PACKET RESERVATION MULTIPLE ACCESS

The R-ALOHA protocol was modified in PRMA to include a centralized base station which broadcasts the status of each time slot [GOOD89] to users in the previous frame. The protocol supports a mixture of data and voice packets with the time frame matched to the periodic rate of voice packets. The packet type is identified using a single bit in the packet header.

Since users are informed of the status of each time slot, it is not necessary for all users to agree on which slot is the first in the frame (i.e., frame boundaries need not be defined in the same manner as R-ALOHA). As in R-ALOHA, users employ the slotted ALOHA protocol to contend for unreserved time slots. Users experiencing collision retransmit after a random delay of q. At the end of each slot, the base station broadcasts the outcome in an acknowledgment packet.

When the base station acknowledges the correct reception of a voice packet, the user reserves the same time slot for future transmission. All other users refrain from using that slot in future frames. When the user stops sending voice packets in the reserved slot, the base station broadcasts this event in an acknowledgment packet. All users are then free to contend for that slot in future frames. Users are allowed to reserve more than one time slot.

A user transmits data packets in unreserved time slots. When a data packet is successfully transmitted, the user does not obtain a time slot reservation. In the event of a collision, data packets are retransmitted with a probability of r. By setting $q > r$, the system gives priority to voice packets.

PRMA is able to achieve only 70% of the capacity with perfect capture [GOOD89] which suggests that the power capture effect does not result in improvements in capacity similar to that of ALOHA. Minislotted PRMA protocol was proposed in [MITR90]. The number of reservation minislots is variable and a minimum bandwidth is reserved for these slots.

7.5 DYNAMIC RESERVATION

The dynamic reservation scheme requires users to make explicit reservations [ROBE73]. A frame comprises a set of fixed-length time slots, with one of these slots further subdivided into reservation minislots. A ready user broadcasts the number of data slots it requires in one of the minislots. If the request is successful, the required number of slots is reserved in the next frame (Figure 7.2). All users must keep track of the queue length (i.e., number of reserved slots) so that when any user makes a successful reservation, it will know how many slots to skip before transmitting. The identity of the users in the queue need not be known. When the reservation queue is empty, all slots in the frame become minislots. This allows a reservation to be transmitted quickly under low traffic load.

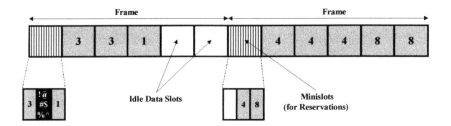

Figure 7.2: Operation of dynamic reservation

7.5.1 Implementation Considerations

In order for all users to keep track of the queue length, the header of each data packet contains information regarding the status of the queue. Alternatively, the queue length may also be announced periodically by a central controller. This information allows new or currently inactive users to join the queue. If the protocol operates in an environment with long propagation delays (e.g., in a satellite environment), a user must update the queue length information with requests received within the channel propagation time just prior to receiving the queue length information. In addition, it is important that each reservation packet be correctly received by all users. Otherwise some collisions will result.

7.5.2 Performance Considerations

If n is the number of data slots in a frame (each of length T_d) and one additional data slot is divided into reservation minislots (each of length T_r), then for an average overall arrival rate of λ, the following stability conditions apply:

$$(n+1)\lambda T_r \leq \frac{1}{e}$$

$$(7.5)$$

$$\lambda T_d \leq \frac{n}{n+1}$$

$$(7.6)$$

Suppose S denotes the throughput. In this case, S represents the fraction of the channel that is reserved. Let R be the fraction of the channel that is idle. The idle portion of the channel is therefore available for random access. Denoting the throughput of the reservation channel as S_r, the normalized rate of successful reservation requests (α) is:

$$\alpha = S_r R = S_r (1 - S)$$

$$(7.7)$$

If v is the ratio of the message length to the reservation bits, then:

$$S = S_r (1 - S) v$$

$$(7.8)$$

$$\therefore S = \frac{S_r}{1/v + S_r}$$

$$(7.9)$$

Equation 7.9 is the same expression obtained for R-ALOHA except that now, the value of v is usually very large. Since v is assumed to be a fixed parameter, the maximum throughput is also a fixed value that is strictly less than one.

7.6 DYNAMIC RESERVATION MULTIPLE ACCESS

In DRMA, the number of reservation slots and their positions change from time to time [LI95]. Any idle time slot can serve as a series of reservation minislots (Figure 7.3). Users with successful reservations in these minislots are informed by the base station at the end of the idle slot. This protocol preserves the characteristics of PRMA but has a smaller probability of collisions due to the use of minislots. Note that one or more idle slot will always exist in this protocol.

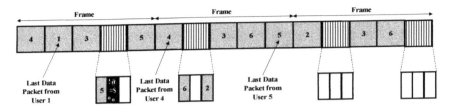

Figure 7.3: Operation of dynamic reservation multiple access

7.7 ROUND-ROBIN RESERVATION

The round-robin reservation scheme requires the number of users to be less than or equal to the total number of fixed-length data slots in a frame [BIND75]. Hence, this scheme requires a fixed number of users unlike the R-ALOHA scheme which will work with unknown or dynamically varying number of users. Each user owns a specific data slot. Whenever there are any extra slots or if the owner of a slot has no message to send, these slots are available for use by other users in a round-robin manner. The owner can reclaim its slot by deliberately generating a collision which causes other users to defer.

To allow other users to know the current state (busy or idle) of an assigned time slot, a user sends its current queue length information in the header of each data packet transmitted in the slot. A zero count indicates that the slot is free. Thus, this scheme is a combination of both implicit and explicit reservation. A user that uses its own slot implicitly reserves it for the next frame. By broadcasting its queue length, a user is explicitly reserving future time slots. Priority can be implemented by having some users own two slots.

7.8 SPLIT-CHANNEL RESERVATION MULTIPLE ACCESS

SRMA is a centrally controlled scheme that divides the available bandwidth into two frequency channels, one for transmitting control information and the second for messages [TOBA76]. Several operational modes are possible. In the request/answer-to-request/message (RAM) mode, the control channel is further divided into two subchannels namely the request subchannel and the answer-to-request subchannel. The request subchannel is operated in the random access mode. If a request packet is correctly received, the controller computes the time at which the backlog on the message channel will empty and then transmits this information in a response packet addressed to the user on the answer-to-request channel.

SRMA, when operated in the RM mode, consists of having only two channels namely, the request channel and the message channel. A successfully transmitted request packet joins the request queue. When the message channel is available, the controller sends a response packet on the message channel. The response packet is addressed to the user scheduled for transmission. If a user does not hear its own address repeated by the controller on the message channel within a certain time after the request is sent, it assumes a collision has occurred and retransmits the request packet.

7.9 INTEGRATED MULTIPLE ACCESS

The integrated multiple access scheme is designed to deal with a mixture of periodic and bursty traffic. Each frame is divided into three subframes [SUDA83]. The reservation subframe comprises a set of minislots. The unreserved subframe is made up of data slots that users contend for using slotted ALOHA. The reserved subframe consists of data slots that are reserved for periodic traffic.

To acquire slots in the reserved frame, a user transmits a reservation in a minislot using slotted ALOHA. If a reservation is successfully transmitted, and if at least one unreserved slot is available in the reserved subframe, the controller sends a confirmation that indicates the position of the newly reserved slot within the reserved subframe. Upon receipt of the confirmation, the user can use the reserved slot in each succeeding frame until it transmits an end-of-message flag. This informs the controller to release the slot for future use.

7.10 DEMAND ACCESS MULTIPLE ACCESS

DAMA is commonly employed in satellite systems. A separate request channel is used by individual users to reserve for capacity when needed. Request packets can be sent using the ALOHA protocol, as is done in INMARSAT DAMA. Alternatively, the request channel can be divided into small, fixed allocation subchannels, which are then assigned to each user. Clearly, the number of users determines the type of access scheme to be used in the request channel. In the INTELSAT SPADE DAMA system, the request channel can be accessed as many as 50 individual users using a 50 ms TDMA frame.

7.11 PRIORITY ORIENTED DEMAND ASSIGNMENT

The PODA reservation scheme uses a movable boundary between the reservation and data subframes that varies with demand [JACO77]. The total frame length is fixed but the reservation subframe changes in length according to the network loading. Two strategies are possible for allocating reservation slots: fixed assignment (FPODA) or contention assignment (CPODA). Since the protocol caters for both data and voice traffic, channel scheduling is not done on a first-come-first-served basis as in the dynamic reservation protocol. In addition to using the reservation subframe, a reservation request can also be piggybacked into the header of a scheduled packet. Thus, only new reservations are sent in the reservation subframe.

A similar scheme using a movable boundary between voice and data sections was described and analyzed in [WIES95]. Data traffic is allowed to use any idle slots of the voice section, resulting in higher utilization. However, voice traffic is not permitted to use unused data slots. This is because a voice connection implies a long term commitment whereas data traffic can be considered short term.

Other sophisticated, movable-boundary TDMA schemes have been suggested to accommodate periodic traffic sources such as voice and video. Although some of these protocols do offer recurring time slots to all users, the recurrence interval is restricted by the length of the frame format rather than being selected to match the desired repetition rate of the traffic serviced. Furthermore, long frames are necessary to optimize delay at maximum network load. This increases the minimum time between successful transmissions, even at light traffic load.

7.12 ANNOUNCED RETRANSMISSION RANDOM ACCESS

An interesting variation of the reservation idea is to use the reservation slots not for reserving packet transmission time but for retransmission in the event of collisions [RAYC85]. The scheme employs a separate control channel to allow a user to announce its intention to retransmit in the event of collision. If a message collides but its reservation survives, then each user knows when the retransmission will be scheduled and avoids using that slot for subsequent messages. The advantage of this scheme is that no additional delay (i.e., to first reserve a data slot before actual data transmission) is required for new packets. Unlike both the binary tree and splitting algorithms, which prevent collisions between new and backlogged by restricting the new arrivals to join the collision resolution interval, ARRA exhibits more flexibility.

7.13 MINISLOTTED PROTOCOL

The operation of the minislotted protocol [KLEI80] is similar to implicit token passing. In this case, an idle time slot is treated as a permission to transmit. The broadcast channel is divided into minislots interleaved with data slots. Each slot is preceded by N minislots, where N is the size of the user population.

The minislotted protocol is useful only for networks that are geographically limited since each minislot has a duration equal or greater than the maximum propagation delay. Suppose v is the ratio of the minislot to data slot duration. Since there are N minislots associated with every data slot, the maximum channel throughput is:

$$S_{max} = \frac{1}{1 + Nv}$$

(7.10)

Clearly, the performance of the minislotted protocol is acceptable only if $Nv \ll 1$. To overcome this limitation, the assigned-slot listen-before-transmission protocol allows a group of users to share a common pool of minislots. There is a tradeoff between the time wasted in collisions and the amount of control overhead required. Channel time is divided into frames, each containing an equal number of minislots, M. Each minislot is assigned a subset of N/M users where N is the total number of users. A ready user senses the channel in its assigned minislot. If the channel is sensed idle, transmission takes place. Otherwise, the packet is rescheduled for transmission in a future frame. The parameter N/M is adjusted according to the load placed on the network. Clearly, for heavy loads, N/M should be unity and the scheme becomes conflict-free.

7.14 BIT-MAP ACCESS PROTOCOL

In BMAP [KNOT85], each user in a network of N users is assigned a unique address from 0 to $N - 1$. Each reservation period is divided into N minislots. For a bit-synchronous channel, a single bit is sufficient for a user to indicate its status (ready or idle). Thus, if user 0 has a packet to transmit, it sends a single pulse on the first minislot. User N announces its intention to transmit on the $N - 1$ slot. After N minislots have passed, all users will have knowledge on which users have data to transmit (i.e., all ready users will be identified after N minislots). The ready users then transmit in numerical order without conflict (Figure 7.4).

The fixed assignment of the minislots in BMAP avoids collisions and removes the need to transmit a user's identity, thereby reducing the duration of these slots. However, higher numbered users encounter longer reservation delays. Furthermore, a user must always wait until the reservation period is over before initiating data transmission.

7.15 BROADCAST RECOGNITION ACCESS METHOD

BRAM [CHLA79] can be viewed as using an imaginary token to implement a form of implicit reservation. This protocol is the same as the basic operation of minislotted alternating priorities (MSAP) [KLEI80] which was discovered independently. BRAM attempts to overcome two main drawbacks of BMAP. Like BMAP, the order in which the users are granted transmission rights is pre-assigned using the numerical value of their addresses. Unlike the BMAP, however, a ready user transmits immediately after its assigned minislot (Figure 7.5). If the user has nothing to transmit, it remains silent for a duration equal to the propagation delay so that transmission rights are implicitly passed on to the next user in sequence.

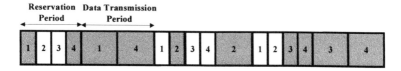

Figure 7.4: Operation of the bit map access protocol (*N* = 4)

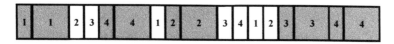

Figure 7.5: Operation of BRAM (*N* = 4)

7.16 MULTILEVEL MULTIPLE ACCESS

The performance of BRAM can be improved using a multi-level, multiaccess protocol (MLMA) [ROTH77]. The maximum throughput efficiency of MLMA is close to BRAM and MSAP but the protocol has a shorter delay under conditions of low network load. As illustrated in Figure 7.6, when a user detects a pulse (i.e., a binary 1) overwriting the most significant bit of its address that is 0, it drops out.

The multilevel address structure substantially reduces the number of bits required for identifying ready users. For example, a population of N users requires only $\log_2 N$ bits for identifying all ready users. MLMA can be modified to give higher priority transmission rights to users that are silent for a long period of time [MOK79]. A user that has most recently transmitted a packet is placed at the bottom of the priority queue.

SUMMARY

Reservation protocols represent a compromise between contention and fixed assignment protocols. Contention protocols are well suited for handling bursty traffic. On the other hand, when user traffic is highly predictable, fixed assignment protocols are usually adopted. Reservation protocols typically cater for users with slowly varying traffic. They differ by the way reservations are made. For a small user population, fixed assignment reservation schemes are attractive while contention reservation schemes are normally deployed in networks with a large number of users.

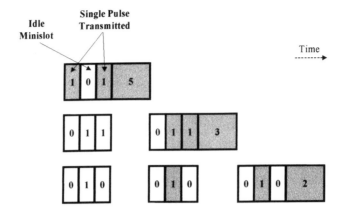

Figure 7.6: Operation of MLMA

BIBLIOGRAPHY

[BALA79]   Balagangadhar, M. and Pickholtz, R., "Analysis of a Reservation Multiple Access Technique for Data Transmission via Satellites", *IEEE Transactions on Communications*, Vol. COM-32, No. 7, July 1984, pp. 1467 – 1475.

[BIND75a] Binder, R., "A Dynamic Packet Switching System for Satellite Broadcast Channels", *Proceedings of the ICC*, June 1975, pp. 41-1 – 41-5.

[BIND75b] Binder, R., McQuillan, J. and Rettberg, R., "The Impact of Multi-access Satellites on Packet Swicthing Networks", *IEEE EASCON*, 1975, pp. 63-A – 63-F.

[BOSE80]   Bose, S., and Rappaport, S., "High Capacity: Low Delay Packet Broadcast Multiaccess", *IEEE Transactions on Aerospace and Electronic Systems*, Vol. AES-16, No. 6, November 1980, pp. 830 – 838.

[CHEN98]   Chen, K., "Multiple Access for Wireless Packet Networks", *Proceedings of the Third Workshop on Multiaccess, Mobility and Teletraffic*, appearing in [EVER98], pp. 327 – 341.

[CHLA79]   Chlamtac, I., *Multiaccess Channel Packet Switching Protocols*, Ph.D Dissertation, University of Minnesota, 1979.

[CHU83]   Chu, W. and Chao, M., "An Analysis of the C-PODA Protocol for a Satellite Data Communication Channel", *Computer and Electronic Engineering*, Vol. 10, No. 3, 1983, pp. 209 – 227.

[CRIS95]   Crisler, K. and Needham, M., "Throughput Analysis of Reservation ALOHA Multiple Access", *Electronics Letters*, January 1995, pp. 87 – 89.

[CROW73]   Crowther, W., Rettberg, R., Walden, D., Ornstein, S. and Heart, F., "A System for Broadcast Communication: Reservation ALOHA", *Proceedings of the 6th Hawaii International Conference on System Sciences*, January 1973, pp. 371 – 374.

[EVER98]   Everitt, D. and Rumsewicz, M., *Multiaccess, Mobility and Teletraffic Advances in Wireless Networks*, Kluwer Academic Publishers, 1998.

[GOOD89]   Goodman, D., Valenzuela, R., Gayliard, K. and Ramamurthi, B., "Packet Reservation Multiple Access for Local Wireless Communications", *IEEE Transactions on Communications*, Vol. 37, No. 8, August 1989, pp. 885 – 890.

[GOOD90]   Goodman, D., "Cellular Packet Communications", *IEEE Transactions on Communications*, Vol. 38, No. 8, August 1990, pp. 1272 – 1280.

[GOOD91]   Goodman, D. and Wei, S., "Efficiency of Packet Reservation Multiple Access", *IEEE Transactions on Vehicular Technology*, February 1991.

[GREE81]   Greene, E. and Ephremides, A., "Distributed Reservation Control Protocols for Random Access Broadcasting Channels", *IEEE Transactions on Communications*, Vol. COM-29, No. 5, May 1981, pp. 726 – 735.

[GRIM75]   Grimsdale, R. and Kuo, F., *Computer Communication Networks* NATO Advanced Study Institutes Series, Sijthoff and Noordhoff International Publishers, The Netherlands, 1975.

[GUHA82]   Guha, D., Schilling, D. and Saadawi, T., "Dynamic Reservation Multiple Access Technique for Data Transmission Via Satellites", *Proceedings of the IEEE INFOCOM*, 1982, pp. 53 – 61.

[JACO77]   Jacobs, I., et al., "CPODA - A Demand Assignment Protocol for SATNET", *Proceedings of the 5th Data Communication Symposium*, September 1977.

[KHAN85]   Khanna, A., *An Analysis of Multiaccess Reservation Strategies for Satellite Channels*, Masters Dissertation, MIT, 1985.

[KLEI78]   Kleinrock, L. and Gerla, M., "On the Measured Performance of Packet Satellite Access Schemes", *Proceedings of the ICCC*, 1978, pp. 535 – 542.

[KLEI80]   Kleinrock, L. and Scholl, M., "Packet Switching in Radio Channels: New Conflict-free Multiple Access Schemes", *IEEE Transactions on Communications*, Vol. COM-28, No. 7, July 1980, pp. 1015 – 1029.

[KNOT85]   Knott, J., "A Fairness Evaluation of the Bit-Map Access Protocol", *IEEE Globecom*, 1985, pp. 34.4.1 – 34.4.3.

[LAM80]   Lam, S., "Packet Broadcast Networks – A Performance Analysis of the R-ALOHA Protocol", *IEEE Transactions on Computers*, Vol. C-29, No. 4, July 1980, pp. 596 – 603.

[LEE84]   Lee, H. and Mark, J., "Combined Random/Reservation Access for Packet-Switched Transmission Over a Satellite", *IEEE Transactions on Communications*, Vol. COM-32, October 1984, pp. 1093 – 1104.

[LI84]   Li, V., "An Integrated Voice and Data Multiple-Access Scheme for a Land-Mobile Satellite System", *Proceedings of the IEEE*, Vol. 72, No. 11, November 1984, pp. 1611 – 1619.

[MARK78]   Mark, J., "Global Scheduling Approach to Conflict-Free Multiaccess via a Data Bus", *IEEE Transactions on Communications*, Vol. COM-26, No. 9, September 1978, pp. 1342 – 1350.

[MARK80]   Mark, J., "Distributed Scheduling Conflict-Free Multiple Access for Local Area Communication Networks", *IEEE Transactions on Communications*, Vol. COM-28, No. 12, December 1980, pp. 1968 – 1975.

[MEDI83]   Meditch, J. and Chu, W., *Computer Communication Networks*, Pergamon Press, 1983.

[MIKK98]   Mikkonen, J., Aldis, J., Awater, G., Lunn, A. and Hutchison, D., "The Magic Wand – Functional Overview", *IEEE Journal on Selected Areas in Communications*, Vol. 16, No. 6, August 1998, pp. 953 – 972.

[MITR90]   Mitrou, N., Orinos, D. and Protonotarios, E., "A Reservation Multiple Access Protocol for Microcellular Mobile Communication Systems", *IEEE Transactions on Vehicular Technology*, November 1990.

[MOK79]   Mok, A. and Ward, S., "Distributed Broadcast Channel Access", *Computer Networks*, Vol. 3, November 1979, pp. 327 – 335.

[NEED95]   Needham, M. and Crisler, K., "QCRA – A Packet Data Multiple Access Protocol for ESMR", *Proceedings of the 45th IEEE Vehicular Technology Conference*, 1995, pp. 336 – 340.

[NG78]   Ng, S. and Mark, J., "A Multiaccess Model for Packet
            Switching with a Satellite having Processing Capability", *IEEE
            Transactions on Communications*, Vol. COM-26 No. 2, February
            1978.

[PAVE86]   Pavey, C., Rice, R. and Cummins, E., "A Performance
            Evaluation of the PDAMA Satellite Access Protocol",
            Proceedings of the IEEE INFOCOM, 1986, pp. 580 – 589.

[PRON98]   Pronk, V., Grinsven, P. and van Driel, C., "A Performance
            Analysis of the Bit-Map Access Protocol for Shared-Medium
            Networks", *1998 International Zurich Seminar on Broadband
            Communications*, pp. 69 – 73.

[QIU96a]   Qiu, X. and Li, V., "Dynamic Reservation Multiple Access
            (DRMA): A New Multiple Access Scheme for Personal
            Communication System (PCS)", *ACM/Baltzer Journal on
            Wireless Networks*, Vol. 2, 1996, pp. 117 – 128.

[QIU96b]   Qiu, X., Li, V. and Ju, J., "A Multiple Access Scheme for
            Multimedia Traffic in Wireless ATM", *Baltzer Journal on Mobile
            Networks and Applications*, Vol. 1, 1996, pp. 259 – 272.

[RAPP79]   Rappaport, S., "Demand Assigned Multiple Access Systems
            Using Collision Type Request Channels: Traffic Capacity
            Comparisons", *IEEE Transactions on Communications*, Vol.
            COM-27, No. 9, September 1979, pp. 1325 – 1331.

[RAYC85]   Raychaudhuri, D., "Announced Retransmission Random
            Access Protocols", *IEEE Transactions on Communications*, Vol.
            33, No. 11, November 1985, 1183 – 1190.

[ROBE73]   Roberts, L., "Dynamic Allocation of Satellite Capacity
            Through Packet Reservations", *Proceedings of the AFIPS
            Conference*, Vol. 42, 1973, pp. 711 – 716.

[ROTH77]   Rothauser, E. and Wild, D., "MLMA - A Collision-Free Multi-
            Access Method", *Proceedings of the IFIP Congress*, 1977, pp. 431 –
            436.

[RUBI79]   Rubin, I., "Access Control Disciplines for Multi-access
            Communication Channels: Reservation and TDMA Schemes",
            IEEE Transactions on Information Theory, Vol. IT-25, No. 5,
            September 1979, pp. 516 – 536.

[SCHO79]   Scholl, M., "On a Mixed Mode Multiple Access Scheme for
            Packet-Switched Radio Channels", *IEEE Transactions on
            Communications*, Vol. COM-27, No. 6, June 1979, pp. 906 – 911.

[SUDA83]   Suda, T., Miyahara, H. and Hasegawa, T., "Performance
           Evaluation of an Integrated Access Scheme in a Satellite
           Communication Channel", *IEEE Journal on Selected Areas in
           Communications*, July 1983.

[SZPA78]   Szpankowski, W., Ono, K. and Urano, Y., "Simulation and
           Analysis of Satellite Packet Switching Computer Networks",
           Proceedings of the ICCC, September 1978, pp. 609 - 616.

[SZAP83]   Szpankowski, W., "Analysis and Stability Consideration in a
           Reservation Multiaccess System", *IEEE Transactions on
           Communications*, Vol. COM-31, No. 5, May 1983, pp. 684 – 690.

[TOBA76]   Tobagi, F. and Kleinrock, L., "Packet Switching in Radio
           Channels: Part III – Polling and (Dynamic) Split-Channel
           Reservation Multiple Access", *IEEE Transactions on
           Communications*, Vol. COM-24, No. 8., October 1976, pp. 832 –
           844.

[TOWS87]   Towsley, D. and Vales, P., "Announced Arrival Random
           Access Protocols", *IEEE Transactions on Communications*, Vol.
           COM-35, No. 5, May 1987, pp. 513 – 519.

[WHIT96]   Whitehead, J., "Distributed Packet Dynamic Resource
           Allocation (DRA) for Wireless Networks", *Proceedings of the 45th
           IEEE Vehicular Technology Conference*, 1996, pp. 111 – 115.

[WIES80]   Wieselthier, J. and Ephremides, A., "A New Class of Protocols
           for Multiple Access in Satellite Networks", *IEEE Transactions
           on Automatic Control*, October 1980, pp. 865 – 879.

[WIES95]   Wieselthier, J. and Ephremides, A., "Fixed and Movable
           Channel Access Schemes for Integrated Voice/Data Wireless
           Networks", *IEEE Transactions on Communications*, Vol. 43, No.
           1, January 1995, pp. 64 – 74.

[WOLE87]   Wolejsza, C., Taylor, D., Grossman, M. and Osborne, W.,
           "Multiple Access Protocols for Data Communications via
           VSAT Networks", *IEEE Communications Magazine*, Vol. 25, No.
           7, pp. 30 – 39.

[WONG91]   Wong, E. and Yum, T., "A Controlled Multiaccess Protocol
           for Packet Satellite Communication", *IEEE Transactions on
           Communications*, Vol. 39, No. 7, July 1991, pp. 1133 – 1140.

Chapter 8

BROADBAND WIRELESS ACCESS PROTOCOLS

Broadband wireless access protocols must adequately address the combined requirements of wireless and multimedia communications. On one hand, the protocol must allow users to share the limited bandwidth resource efficiently. This implies two criteria: maximizing the utilization of the radio frequency spectrum and minimizing the delay experienced by the users. On the other hand, the multiple access protocol operating in multimedia networks is required to handle a wide range of information bit rates together with various types of real-time and nonreal-time service classes with different traffic characteristics and quality of service guarantees.

8.1 THE 2.4 GHz IEEE 802.11 WIRELESS LAN STANDARD

The 2.4 GHz IEEE 802.11 standard [IEEE97], [IEEE99b] specifies wireless connectivity for fixed, portable, and moving users in a geographically limited area. The supported data rates are listed in Table 8.1.

The 802.11 physical layer specification allows three transmission options namely:

❑   direct-sequence spread spectrum;
❑   frequency-hopping spread spectrum;
❑   diffuse infrared.

Table 8.1: IEEE 802.11 data rate specifications

Data rate (Mbit/s)	Symbol rate (Msymbols/s)	Bits per symbol	Code length	Modu-lation
1	1	1	11-bit Barker Code	DBPSK
2	1	2	11-bit Barker Code	DQPSK
5.5	1.375	4	8-bit CCK	DQPSK
11	1.375	8	8-bit CCK	DQPSK

Note that spread spectrum is not used as a multiple access technique in 802.11 wireless LANs. Rather, it is used to protect data signals against the effects of multipath propagation and other impairments [BING00].

The standard defines two medium access control (MAC) protocols. The distributed coordination function (DCF) employs CSMA with collision avoidance (CSMA/CA) for contention-based multiple access. Contention-free service is provided by the point co-ordination function (PCF) which is essentially a polling access method. Unlike the DCF, the implementation of the PCF is not mandatory. The PCF relies on the asynchronous access service provided by the DCF.

8.1.1 The Distributed Coordination Function

According to the DCF, when a user decides to transmit a new packet, it must first sense the channel. If the channel is sensed to be idle for a duration greater than the distributed interframe space (DIFS), the packet is transmitted immediately. Otherwise the transmission is deferred and the backoff process is initiated. The backoff process is also initiated immediately after a successful packet transmission. The backoff process is a collision avoidance technique that is performed to reduce the high probability of collision. It introduces a random interframe space between successive packet transmissions. Thus, the backoff process is effectively an attempt to separate the total number of ready users into smaller groups, each using a different time slot.

If the channel is detected to be busy, the user must delay until the end of the DIFS interval and further wait for a random number of time slots (called the backoff interval) before attempting transmission (Figure 8.1). Specifically, a user computes a backoff interval that is uniformly distributed between zero and a maximum contention window. This backoff interval is used to initialize the backoff timer. The timer is decremented only when the channel is idle but is suspended when another user is transmitting. When the channel is idle, the user waits for a DIFS and then decrements the backoff timer periodically. The decrementing period is called the slot time, which corresponds to the maximum round-trip propagation delay between two users in the same radio coverage area.

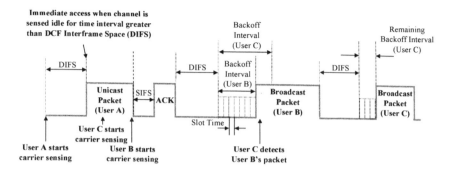

Figure 8.1: Broadcast and unicast packet transmissions using CSMA/CA

At each slot time, carrier sensing is performed to determine if there is activity on the channel. If the channel is idle for the duration of the slot, the backoff interval is decremented by one slot. If a busy channel is detected, the backoff procedure is suspended and the backoff timer will not decrement for that slot. In this case, when the channel becomes idle again for a period greater than the DIFS, the backoff procedure continues decrementing from the slot which was previously disrupted. This implies that the selected backoff interval is now less than the first. Hence, a packet that was delayed while performing the backoff procedure has a higher probability of being transmitted earlier than a newly arrived packet. This ensures some measure of fairness in channel access. The process is repeated until the backoff interval reaches zero and the packet is transmitted.

The 802.11 MAC protocol requires the receiver to send a positive acknowledgment (ACK) back to the transmitter if a packet is received correctly. The ACK is transmitted after the short interframe space (SIFS) which is of a shorter duration than the DIFS. This enables an ACK to be transmitted before any new packet transmission. Thus, a receiving user need not sense the channel before transmitting the ACK. If no ACK is returned, the transmitter assumes the packet is corrupted (either due to a collision or transmission error) and retransmits the packet. Hence, unlike the CSMA/CD protocol used in Ethernet, in CSMA/CA, collisions are only inferred only after the entire data packet is transmitted. Note that without collision detection, packets transmitted using CSMA need not be of a certain minimum length.

To reduce the possibility of repeated collisions, after each unsuccessful transmission attempt, the contention window is doubled until a predefined threshold is reached. Thus, the backoff interval increases exponentially up to a maximum contention window.

8.1.2 Virtual Sensing

The hidden user phenomenon (Section 5.6.3) arises when a user is able to successfully receive packets from two different users but these two users cannot receive signals from each other. In this case, it is possible for one of the two users to sense the channel as being idle even when the other is transmitting. This results a collision at the receiving user.

The DCF deals with this problem by incorporating a virtual carrier sense mechanism that distributes reservation information by announcing the impending use of the wireless channel. The basic idea is a four-way handshake where short control packets called request-to-send (RTS) and clear-to-send (CTS) packets are exchanged prior to the transmission of the data packet (Figure 8.2).

The RTS packet is issued by the source user while the CTS packet is issued by the destination user to grant the source user permission to transmit. On correct receipt of the data packet, the destination user sends an ACK packet to the source user. The RTS and CTS packets contain the source and destination addresses as well as a duration field defining the time interval between the transmission of the data packet and the returning ACK packet.

It is clear that with virtual sensing, collisions occur only during the transmission of a RTS packet. The RTS packet overlaps either with a RTS or a CTS packet from another user. The short RTS and CTS packets minimize the overhead due to such collisions and in addition, allow the transmitting user to infer collisions quickly. Moreover, the CTS packet alerts neighboring users (that are within the range of the receiving user but not of the transmitting user) to refrain from transmitting to the receiving user, thereby reducing hidden user collisions (Figure 8.3). In the same way, the RTS packet protects the transmitting area from collisions when the ACK packet is sent from the receiving user. Thus, reservation information is distributed among users surrounding both the transmitting and receiving users (i.e., users that can hear either the transmitter or the receiver or both).

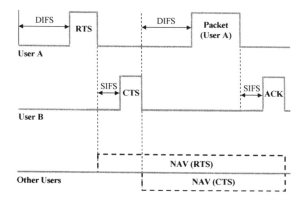

Figure 8.2: Packet transmission using virtual sensing

All non-transmitting users that successfully decode the duration field in the RTS and CTS packets store the channel reservation information in a Net Allocation Vector (NAV). For these users, the NAV is used in conjunction with carrier sensing to detect the availability of the channel. Hence, these users will defer transmission if the NAV is non-zero or if carrier sensing indicates that the channel is busy. Note that the NAV is a timer, which, unlike the backoff timer, is continuously decremented regardless of whether the channel is sensed busy or idle. Since users can hear either the transmitter or the receiver refrain from transmitting until the NAV expires, the probability of collision in a hidden user situation is reduced. Like the ACK mechanism, virtual sensing cannot be applied to data packets with broadcast and multicast addressing due to a high probability of collision among a potentially large number of CTS packets.

Due to the large overhead involved, the mechanism is not always justified, particularly for short data packets. Hence, the 802.11 standard allows short packets to be transmitted without virtual sensing. This is controlled by a parameter called the RTS threshold. Only packets of lengths above the RTS threshold are transmitted using virtual sensing. Note that the effectiveness of the virtual sense algorithm depends strongly on the assumption that both the transmitting and receiving users have similar operating ranges (i.e., transmitter power and receiver sensitivity are about the same). Use of virtual sensing is optional but the mechanism must be implemented.

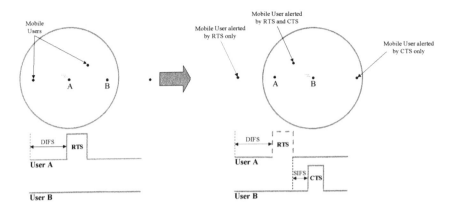

Figure 8.3: Distribution of reservation information in virtual sensing

8.1.3 The Point Coordination Function

The optional PCF may be used to support time-bounded services. PCF employs a centralized access scheme where users in a wireless coverage area are allowed to transmit only when polled by a central controller (called an access point), which has priority access to the channel. To allow other users with asynchronous data to access the channel, the 802.11 MAC protocol alternates between DCF and PCF, with PCF having higher priority access. This is achieved using a superframe concept where the PCF is active in the contention-free period, while the DCF is active in the contention period (see Figure 8.4).

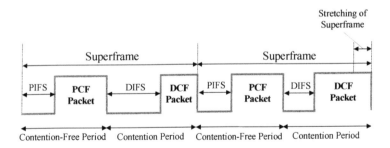

Figure 8.4: Co-existence of PCF and DCF in a superframe

A beacon packet notifies all users in the wireless coverage area to refrain from transmission for the duration of the contention-free period (unless specifically polled by the access point). The contention-free period can be variable in length within each superframe without incurring any additional overhead. At the beginning of the superframe, if the channel is free, the PCF gains control over the channel. If the channel is busy, then PCF defers until the end of the packet or after an acknowledgment has been received. Since the PIFS is of shorter duration compared to DIFS, PCF can gain control of the channel immediately after the completion of a busy period.

Note that the contention period may be of variable lengths and this causes the contention-free period to start at different times. Similarly, a packet may start near the end of the contention period, thereby stretching the superframe and causing the contention-free period to start at different times. Thus, the repetition interval of the contention-free period can vary depending on the network load. Another important observation is that collisions may be introduced when access points transmit polling messages to mobile users stationed within overlapping wireless coverage areas.

8.2 THE 5 GHz IEEE 802.11 WIRELESS LAN STANDARD

The IEEE 802.11 committee has also finalized a 5 GHz standard (approved as IEEE 802.11a on 16th September 1999 [IEEE99a]) that will support wireless data rates of 6 to 54 Mbit/s based on coded OFDM. The main specifications are listed in Table 8.2.

Table 8.2: Specifications for the 5 GHz IEEE 802.11 physical layer

Parameters	Specification
Mandatory data rates (Mbit/s)	6, 12, 24
Optional data rates (Mbit/s)	18, 36, 48, 54
Number of subcarriers	52 (48 for data, 4 for pilots)
Sampling rate	20 Msamples/s
Guard interval	800 ns (16 time samples), 400 ns as option
Channel spacing	20 MHz
Signal bandwidth	16.6 MHz
Modulation for subcarrier	BPSK, QPSK, 16-QAM, 64-QAM
Bit-interleaved convolutional coding	Constraint length = 7, Rate = 1/2, 9/16, 3/4

OFDM has been chosen due to its excellent performance in highly dispersive channels. OFDM also allows considerable flexibility in the choice of different modulation methods. The channel spacing of 20 MHz is a compromise between having high data rates per channel and having a reasonable number of channels in the allocated spectrum.

Out of the 52 subcarriers in each channel, 48 are subcarriers carrying user data while the remaining 4 subcarriers are pilots that facilitate phase tracking for coherent demodulation. Each subcarrier serves as one communication link between the access point and the mobile terminals. The 800 ns guard interval is sufficient to enable good performance on channels with delay spread of up to 250 ns. An optional shorter guard interval of 400 ns may be used in small indoor environments.

BPSK, QPSK, and 16-QAM are the supported subcarrier modulation schemes with 64-QAM as an option. Forward error correction is performed by convolutional coding with rate 1/2 and constraint length of 7. The three code rates of 1/2, 9/16, and 3/4 are obtained by code puncturing.

8.3 THE HIPERLAN TYPE 1 WIRELESS LAN STANDARD

HiperLAN Type 1 is a European standard for 23.5 Mbit/s radio LAN that operates between 5.15 – 5.30 GHz. This band is compatible with the U-NII band in the US. Unlike the 802.11 standard, which supports both the ad-hoc (distributed) and infrastructure (centralized) topologies, HiperLAN Type 1 supports only the ad-hoc topology. However, HiperLAN Type 1 also caters for the multihop ad-hoc topology, as opposed to the single-hop topology adopted by 802.11. This allows HiperLAN Type 1 networks to be implemented without the need for frequency planning.

8.3.1 The HiperLAN Type 1 MAC Protocol

HiperLAN Type 1 is designed primarily for high-speed data traffic although some limited QoS provisions are included. HiperLAN Type 1 employs a fully distributed MAC protocol called Elimination Yield Non Pre-emptive Multiple Access (EY-NPMA). The MAC protocol overcomes the hidden user problem by making use of the fact that a receiving node can sense (detect) a signal even though it is unable to decode the signal (i.e., the sensing range can be much greater than the receiving range [BING00].

EY-NPMA is essentially CSMA with added prioritization. Channel access is non pre-emptive because only data packets ready at the start of each channel access cycle are allowed to contend. Users undergo contention resolution based on priority assertion and random backoff (Figure 8.5). Clearly, if two users have different access patterns, then the listening and transmission periods as well as the eventual packet transmission will not coincide. The slot time is of variable length, depending on whether the slot is a listening period or a data transmission period.

When a user desires transmission and the channel is sensed idle for a minimum duration of 2000 bit times, EY-NPMA allows immediate access. On the other hand, if the channel is not free, then channel access with synchronization takes place. Synchronization is achieved by forcing a channel access cycle to start after the end of the previous packet transmission. The channel access cycle consists of three phases namely:

❑   Prioritization;
❑   Contention;
❑   Transmission.

The prioritization phase comprises five slots of 168 bits each. The packet with the highest priority to transmit corresponds to the packet with holding an access pattern with the largest decimal number. Hence, an access pattern with the highest priority level (i.e., 5) will show no idle slots at the beginning while an access pattern of the lowest priority will have four idle slots.

The contention interval can be further divided into two phases:

❑   Elimination;
❑   Yield.

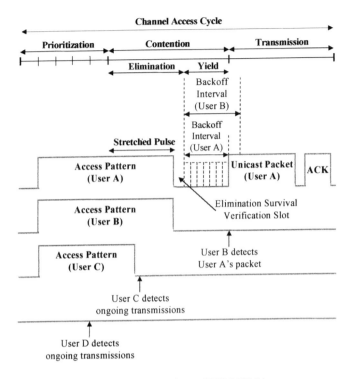

Figure 8.5: Operation of EY-NPMA

The elimination phase involves stretching the pulse with a random number of slots (0 to 12) of 212 bits each. Pulse stretching is performed independently by a transmitting user. The duration of the pulse varies according to a geometric distribution of probability $p = 0.5$. Thus, the pulse is larger than one slot time with a probability of 0.5, greater than two slots with a probability of 0.25, and so on. After the stretched pulse, users perform carrier sensing on an idle slot of 256 bits. The elimination phase ends with this idle slot (called elimination survival verification slot). Only users that simultaneously hold the highest access priority and select the longest stretched pulse survive the elimination phase and proceed to the yield phase. In this phase, a random number of idle slots (0 to 9) is selected according to a truncated geometric probability distribution with parameter $r = 0.1$. The length of each slot is 168 bits. If a user detects no signal after a duration equal to the total number of idle slots that has elapsed, the data packet is transmitted. Otherwise the user defers till the end of the current packet transmission. Note that the prioritization and contention intervals essentially comprise two listening periods separated by one transmission period (the access pattern), each of different lengths. The use of only one transmission period reduces radio switching overhead [BING00]. This explains why the prioritization slots are based on decimal numbers and not binary digits (which will require less slots).

The objectives of the elimination and yield phases are complementary. The elimination phase significantly reduces the initial number of contending users taking part in the channel access cycle. Its performance is largely insensitive to the total number of contending users. The yield phase performs well with a small number of contending users. Like 802.11, unicast packets are explicitly acknowledged while broadcast/multicast packets are not acknowledged.

8.3.2 Quality of Service

HiperLAN defines five channel access priority levels according to residual (useful) lifetime and user priority (Table 8.3). The user priority is an attribute assigned to each packet according to the traffic type it carries. The residual lifetime represents the maximum time interval the packet must be delivered. Since multihop routing is supported, the residual lifetime is normalized to the number of hops it has to travel before reaching the final destination.

The residual lifetime is computed as follows. Whilst a lower priority packet is waiting, its residual lifetime will be decremented. The user may decide to increase the priority of a packet as its residual lifetime decreases. When the residual lifetime becomes zero and the packet has not been serviced, it will be discarded. Within the same priority class, FCFS policy prevails. Hence, the MAC protocol in HiperLAN Type 1 provides either best effort latency for isochronous traffic (e.g., voice, video) or best effort integrity for asynchronous traffic (e.g., data).

8.4 THE HIPERLAN TYPE 2 WIRELESS LAN STANDARD

HiperLAN Type 2 allows wireless LANs to be interconnected to virtually any type of fixed network technology. It can carry Ethernet or IP packets, ATM cells, and supports UMTS. The standard provides wireless data rates of up to 54 Mbit/s at the physical layer and up to 25 Mbit/s at the network layer with QoS guarantees such as maximum allowable delay and cell loss ratio. Like the 5 GHz IEEE 802.11 standard, HiperLAN Type 2 is based on OFDM. The physical layer specifications are similar to those of the 802.11 standard depicted in Table 8.2. A key feature of the physical layer is the provision of several modulation and coding configurations (Table 8.4). This allows a HiperLAN Type 2 network to adapt to changing radio link quality.

Unlike 802.11, HiperLAN Type 2 is connection-oriented. Connections must be established between the mobile terminal and the access point prior to data transmission. This is achieved using signaling functions. The connections are time-division multiplexed over the air interface and can be point-to-point and point-to-multipoint. Point-to-point connections are bidirectional whereas point-to-multipoint connections are unidirectional (towards the mobile terminal).

Table 8.3: Priority and residual lifetime

Normalized residual lifetime (NRL)	High user-defined priority	Low user-defined priority
NRL < 10 ms	0	1
10 ms ≤ NRL < 20 ms	1	2
20 ms ≤ NRL < 40 ms	2	3
40 ms ≤ NRL < 80 ms	3	4
NRL ≥ 80 ms	4	4

Table 8.4: Modulation and coding parameters for HiperLAN Type 2

Data rate	Modulation	Code rate	Coded bits per subcarrier	Data bits per OFDM symbol
6	BPSK	1/2	1	24
9	BPSK	3/4	1	36
12	QPSK	1/2	2	48
18	QPSK	3/4	2	72
27	16-QAM	9/16	4	108
36	16-QAM	3/4	4	144
54	64-QAM	3/4	6	216

8.4.1 HiperLAN Type 2 MAC Protocol

In the HiperLAN Type 2 MAC protocol, the access point exercises centralized control and adapts according to the resources demanded by each mobile terminal. The protocol is based on TDD and dynamic TDMA. TDD allows communication in the downlink and uplink within the same time-slotted frame. The time slots are allocated dynamically depending on the need for bandwidth resources.

8.4.2 HiperLAN Type 2 Frame Format

As illustrated in Figure 8.6, the basic MAC frame structure has a fixed duration of 2 ms and comprises transport channels for:

- broadcast control (BCH);
- frame control (FCH);
- access control (ACH);
- downlink (DL) and uplink (UL) data transmission;
- random access (RCH).

The functions of the transport channels are listed in Table 8.5. Data from both the access point and mobile terminals are transmitted in dedicated time slots, except for the RCH where contention for the same time slot is possible. The results from the RCH access are reported back to the mobile terminals via the ACH. The duration of BCH is fixed whereas the duration of other fields is dynamically adapted to the current traffic situation. For example, as more mobile terminals request for bandwidth resources, the access point will allocate more resources for the RCH.

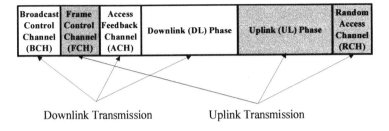

Figure 8.6: Frame format for HiperLAN Type 2

The UL and DL traffic consists of protocol data units (PDUs) to and from the mobile terminals. Each mobile terminal can transmit a series of PDUs if resources have been granted in the FCH. A PDU comprises two sections:

❏   data link layer user PDUs (U-PDUs) of 54 octets with 48 octets of payload;
❏   data link layer control PDUs (C-PDUs) of 9 octets.

8.4.3 QoS Support

The connection-oriented nature of HiperLAN Type 2 allows straightforward implementation of QoS support. Each connection can be assigned a specific QoS in terms of bandwidth, delay, jitter, and bit error rate. It is also possible to employ a simpler approach where each connection is assigned a priority level relative to other connections. The QoS support enables the transmission of a mixture of traffic types of (e.g. voice, video, and data). There are also specific connections for unicast, multicast, and broadcast transmission.

Table 8.5: Functions of the transport channels

Channel	Function
BCH	contains control information such as transmission power levels, start and length of the FCH and RCH, wake-up indicator (for power management), and identifiers (for identifying both the HiperLAN Type 2 network and the access point)
FCH	describes how resources are allocated within the current MAC frame in the DL, UL, and RCH
ACH	contains information on previous access attempts made in the RCH
RCH	used by the mobile terminals to request bandwidth resources for the DL and UL in the upcoming MAC frames, and to exchange signaling messages.

8.5 HOME NETWORKS

The past two years have seen a growing demand for home wireless networks. For instance, the HomeRF Working Group (supported by the ITU) and includes more than 40 leading companies while the Bluetooth Special Interest Group has a current support strength of some 1700 members. Home wireless networks differ from wireless LANs in that they merely provide cable replacement. However, users get integrated voice and data transmission that is highly flexible and affordable. There is no need to string cable between computers and peripherals or attach phones to existing telephone jacks. Connections between computers, peripherals, and handsets are entirely wireless. A summary of HomeRF's specifications is provided in Table 8.6.

8.5.1 HomeRF's Shared Wireless Access Protocol

HomeRF's Shared Wireless Access Protocol (SWAP) specification defines an over-the-air interface that is designed to support both wireless voice and data traffic. SWAP adopts a hybrid access protocol:

❑   TDMA for delivery of interactive voice and other isochronous services;
❑   CSMA/CA for delivery of asynchronous high-speed packet data.

The hybrid TDMA/CSMA frame structure is shown in Figure 8.7.

Table 8.6: Main system parameters for HomeRF

Parameter	Specification
Hopping rate	50 hop/s
Frequency range	2.4 GHz ISM band
Transmit power	100 mW (20 dBm)
Data rate	1 Mbit/s (2-FSK), 2 Mbit/s (4-FSK)
Range	Up to 50 m
Number of users	Up to 127 per network
Voice connections	Up to 6 full duplex connections, with error control

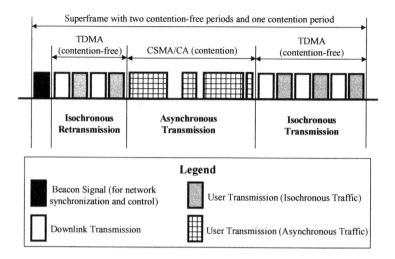

Figure 8.7: HomeRF packet structure

8.5.2 Bluetooth's Access Mechanism

Bluetooth is a wireless data interface standard that provides a simple means of exchanging data between two portable communications devices (e.g., mobile phones, personal computers). Bluetooth operates in the unlicensed 2.4 GHz ISM band. The standard supports two types of connections:

❑   Synchronous Connection Oriented (SCO);
❑   Asynchronous Connectionless (ACL).

SCO packets are transmitted over reserved slots in a point-to-point connection between a controlling master unit and a slave device. Once the connection is established, both master and slave may send SCO packets. A SCO packet allows both voice and data transmission. However, only the data portion is retransmitted when corrupted. The ACL connection supports both symmetric and asymmetric data traffic. The master unit controls the connection bandwidth and decides how much bandwidth is given to each slave. Slaves must be polled before they can transmit data (Figure 8.8).

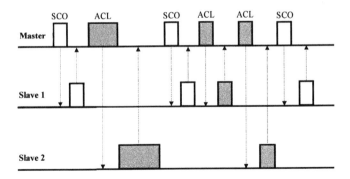

Figure 8.8: Polling mechanism in Bluetooth

Bluetooth can support one asynchronous data connection, up to three synchronous voice connections or a connection which simultaneously supports a mixture asynchronous data and synchronous voice. Each synchronous connection supports a data rate of 64 Kbit/s. An asynchronous connection supports 721 Kbit/s in the forward direction while permitting 57.6 Kbit/s in the reverse direction. Alternatively, it can support a symmetric connection of 432.6 Kbit/s.

8.6 WIRELESS ATM

Many high-speed wireless ATM projects have been proposed recently. The wireless ATM concept combines the user mobility with statistical multiplexing and QoS guarantees provided by ATM networks. It also facilitates deployment of integrated, high-speed wireless access in the local loop. Several wireless ATM access protocols are described in this section.

8.6.1 Simple Asynchronous Multiple Access

SAMA [DELL97] provides a simple bandwidth setup mechanism with support for widely varying data rate requirements. The protocol can be slotted or unslotted. The operation of slotted SAMA is shown in Figure 8.9.

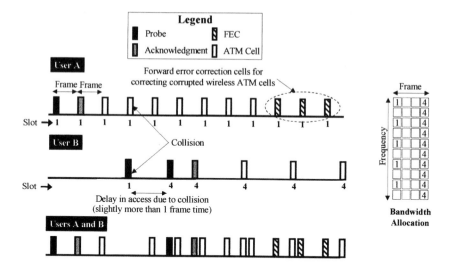

Figure 8.9: Operation of slotted SAMA

8.6.2 Distributed Queuing Request Update Multiple Access

DQRUMA is a demand-assignment multiple access protocol which
provides priority-based scheduling in the downlink and uplink directions. It
was designed for the wireless Broadband Ad-Hoc ATM LAN (BAHAMA),
a self-organizing ad-hoc network developed by Bell Laboratories.

The operation of DQRUMA protocol requires portable base stations to
control the bandwidth allocated to users. Uplink and downlink
communications are performed on different frequency channels using
FDD. Users transmit reservation requests using a contention scheme on the
uplink, which is divided into two time-slotted channels, namely the Request
Access channel and the Packet Transmit channel (Figure 8.10). There is no
frame reference. Request access and packet transmission are all done on a
slot-by-slot basis. Note that handling of acknowledgments is easier to
implement using FDD than TDD.

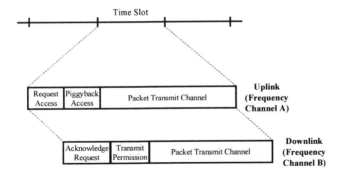

Figure 8.10: Operation of DQRUMA

A unique feature of this protocol is that it allows a user to request additional slots by piggybacking the transmission of ATM cells that are queued in the buffer. Users need to send requests to the base station only for packets that arrive at an empty buffer. This helps to cut down the number of reservations (and access conflicts).

In wireless ATM systems, the physical layer overhead can be substantial relative to the length of an ATM cell. Thus, a question arises as to whether or not it is more efficient to transmit a short request (rather than the entire ATM cell) when making reservations. It turns out that even when a large physical layer overhead is taken into account, it is still better to transmit a short request than to allow contention of full ATM cells [KARO96].

8.6.3 MASCARA

Magic WAND (Wireless ATM Network Demonstrator) [MIKK98] is developed for customer premise networks under the Advanced Communications Technologies and Services (ACTS) program that is funded by the European Union. The project covers a whole range of functionality from basic wireless data transmission to shared multimedia applications.

The multiple access protocol and data link layer control layers for Magic WAND have been combined as one layer, bearing the name of mobile access scheme based on contention and reservation for ATM (MASCARA). MASCARA is a centrally controlled, adaptive TDMA/TDD scheme which combines reservation and contention to achieve efficient transmission and QoS guarantees. The traffic scheduling algorithm is delay-oriented and is designed to meet the requirements of different ATM service classes. A variable frame format is chosen so that the scheduling algorithm can assign time slots to user terminals based on traffic demands (Figure 8.11). Since a wireless link cannot provide a consistent cell loss ratio (CLR), the data link layer adds ARQ capability to sustain the negotiated CLR.

Communication among mobile terminals and base stations takes place in the 5 GHz frequency band at a data rate of 20 Mbit/s and at a maximum range of 50 m. Higher data rate operation at 50 Mbit/s in the 17 GHz frequency band is also being studied. Figure 8.12 illustrates the structure of a single OFDM symbol in Magic WAND. Each symbol period is 1.2 us (13.3 Msymbols/s) and comprises the ramp-up/ramp-down (filter) time as well as guard time. With 16 subcarriers and 8-PSK modulation, this gives a raw data rate of 40 Mbits/s. The spacing of subcarriers is 1.25 MHz, the reciprocal of the FFT period in order for the subcarriers to be orthogonal. Thus, the total bandwidth is 16 × 1.25 MHz or 20 MHz.

8.6.4 WATMnet Access Protocol

WATMnet is developed by the Computer and Communications (C&C) Research Laboratories of NEC in the US [RAYC99]. The access protocol is based on TDMA with TDD and contention (Figure 8.13). Downlink control packets and ATM data cells are multiplexed into a single TDM burst. Uplink control (e.g., user authentication and registration) packets are sent using slotted ALOHA while ATM signaling and data cells are transmitted via TDMA slots assigned by a central controller. Bandwidth allocation is controlled by base stations with mobile terminals scheduling traffic on the allocated bandwidth. The ATM data slots are partitioned into two groups:

❑ rate mode (for ABR, VBR, and CBR traffic);
❑ burst mode (for UBR traffic).

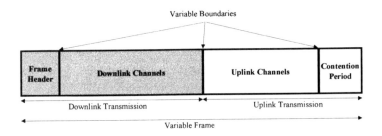

Figure 8.11: Operation of MASCARA

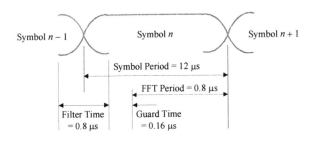

Figure 8.12: OFDM parameters in Magic WAND

UBR/ABR slots are assigned dynamically on a frame-by-frame basis while CBR slots are assigned periodically once a connection is successfully established. For VBR, some fixed slots are assigned together with several extra slots which are allocated on a usage basis. CBR and VBR connections can be blocked while ABR/UBR connections are always accepted subject to appropriate flow control.

8.6.5 AWACS Access Protocol

The Advanced Wireless ATM Communications Systems (AWACS) [UMEH96] is a co-operative project between Europe and Japan. It is closely related to the System for Advanced Mobile Broadband Applications (SAMBA). The major radio parameters of these two projects are listed in Table 8.7.

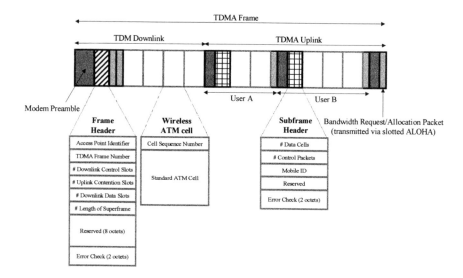

Figure 8.13: Operation of Dynamic TDMA/TDD

Table 8.7: Radio parameters for SAMBA and AWACS

Radio parameters	SAMBA	AWACS
Frequency band (GHz)	40	19
Access method	TDMA/FDD	TDMA/TDD
Wireless bit rate (Mbit/s)	2 × 64	70
Modulation	OQPSK	OQPSK
Radio cell size (m)	6 – 200	50 – 100

The base and mobile stations employ directional antennas and their locations were optimized through ATM cell error measurements. AWACS is based on TDMA/TDD and aimed at allowing both symmetric and asymmetric ATM cell transmission. One of the unique features of AWACS is the reduction in delay spread using narrowbeam directional antennas. This is possible since the system considers low-mobility terminals. The frame format consists of 32 time slots with a period of 32 ms. Each slot can be a transmission in the uplink or downlink.

8.7 SATELLITE ATM

With unobstructed views of virtually the entire world, satellites are poised to deliver interactive broadband services in ways even advanced ground-based networks will be hard to match. The new breed of satellites will act as powerful signal repeaters in the sky, receiving and resending radio transmissions from ground-based antennas.

8.7.1 Multibeam Satellite Systems

Multibeam satellite systems are capable of frequency reuse over fixed terrestrial radio cells to ensure efficient use of the spectrum and as a result, can support large capacities. The cells are scanned in a regular manner by the satellite's transmit and receive beams, resulting in TDMA among the radio cells. Each earth station locks on a different satellite beam so they do not interfere with each other. The traffic streams from individual earth stations are ordered in a TDMA frame structure (Figure 8.14).

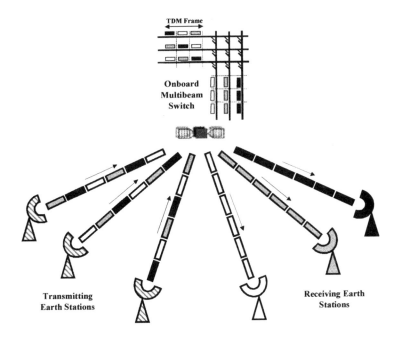

Figure 8.14: A multibeam TDMA satellite system

8.7.2 Multifrequency TDMA

A popular method for multiple access in satellite ATM systems is bit synchronous multifrequency TDMA (MF-TDMA) as depicted in Figure 8.15. The access technique allows TDMA to be operated without any guard band and with low connection blocking.

MF-TDMA employs smaller satellite antennas that make efficient use of transmission power. Each earth station may transmit on any frequency channel at a given time. The MF-TDMA frame is divided into two parts, each with a set of fixed-length slots. The signaling and synchronization slots allow user terminals to request and receive both signaling and timing information needed for registration, connection establishment and synchronization. ATM cells are transmitted in the data slots. Each data slot accommodates an ATM cell with extended header information such as forward error correction coding and in-band signaling.

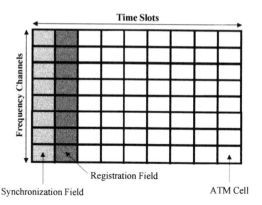

Figure 8.15: A typical multifrequency TDMA frame

MF-TDMA allocates bandwidth on-demand, thus ensuring efficient use of satellite resources according to actual user needs. Since signaling information must be exchange regularly between the ground terminal and the satellite, the signaling and synchronization sections are based on fixed assignment [HUNG98]. ATM data connections are however, based on demand assignment.

There are two basic types of demand assignment namely:

❑   fixed-rate;
❑   variable-rate.

With fixed-rate assignment, a connection is allocated a fixed number of slots, either on a per frame basis or a multiframe basis. If the source is idle, the assigned slots will be wasted. However, this wastage is only for the duration of the connection and not permanent. Variable-rate demand assignment further reduces under-utilization of bandwidth resources by assigning slots only when it is known that there are ATM cells awaiting service at the terminal's queue. A drawback of this scheme, however, is the need to transmit signaling information about the status of the terminal's queue and this incurs additional propagation delay.

8.8 WIRELESS LOCAL LOOP

The WLL allows a long-distance carrier to bypass the local service provider, thereby cutting down subscriber costs. Local Multipoint Distribution Service (LMDS) and the Multichannel Multipoint Distribution Service (MMDS) are two major fixed WLL solutions for delivery of broadband services to customer premises (e.g., residential homes, business offices) and are seen as strong competitors to wireline alternatives (e.g., ADSL, cable modems). Both MMDS and LMDS employ cellular architectures. However, unlike mobile phone networks, the links between the central hub and the distributed users within each radio cell are fixed. Typical cell coverage areas for MMDS and LMDS range from 25 to 35 miles and 1 to 5 miles respectively. Standardization activities are currently being undertaken by the ATM Forum, DAVIC, ETSI, and ITU. The majority of these activities focus on ATM as the transport mechanism.

8.8.1 MMDS

MMDS operates with a bandwidth of 500 MHz in the 2.150 – 2.682 GHz band and provides large capacities in the order of 10's of Mbit/s (a potential capacity of around 200 video channels).

8.8.2 LMDS

LMDS was originally intended for consumer services with limited interactivity (e.g., digital TV broadcasting, video-on-demand). It was later recognized that LMDS systems have a strong potential to supply broadband services to both homes and businesses and the interest gradually shifted towards these applications.

LMDS typically operates at Ka millimeter wave band (28 to 31 GHz) and extremely large blocks of allocated spectrum of approximately 1 GHz are available. Thus LMDS promises much larger capacities in the order of 100's of Mbit/s data rates on each link and is capable of supporting emerging broadband telecommunication services including fast Internet access, digital video distribution, video teleconferencing, and other interactive switched multimedia services.

Although channel impairments play a significant role in the design of both MMDS and LMDS systems, LMDS requires special attention since millimeter waves are very much affected by outages due to rain. Thus, the implementation of LMDS demands many innovations in modulation, channel coding, and adaptive antenna techniques.

8.8.3 Access Methods in LMDS

The customer premise equipment can be attached to LMDS networks using TDMA, FDMA, or CDMA. Currently, TDMA and FDMA are the predominant access methods. These methods apply only to uplink transmissions from the customer site to the hub. The downlink traffic from the hub is based on time-division multiplex.

8.9 IMT-2000

Third generation (3G) global wireless technologies will become available over a period of several years, starting from 2000 in some countries and as late as 2002 in others. The governing standard is ITU's International Mobile Telephony 2000 or IMT-2000 (previously known as FPLMTS). This air-interface standard, provides worldwide compatibility among wireless systems and is supported by TIA in the US, ETSI in Europe, and ARIB in

Asia. IMT-2000 provides a total bandwidth of 230 MHz in the 1885 – 2025 MHz and 2110 – 2200 MHz bands that has been identified by the World Administrative Radio Conference in 1992 (WARC '92).

ITU's vision of global wireless access in the 21st century, including mobile and fixed access, is aimed at providing directions to the diverse second generation (2G) technologies in the hope of unifying these competing wireless systems into a seamless, 3G radio infrastructure capable of offering a wide range of services (Figure 8.16). The benefits are enormous since a successful IMT-2000 standard creates a single market for all aspects of cellular telephony.

8.9.1 Coverage Areas and Data Rates

IMT-2000 aims to provide ubiquitous wireless communications in many different environments. It covers both terrestrial and satellite networks, from indoor "pico" radio cells through outdoor "micro" and "macro" cells to satellite "mega" cells. In addition to international roaming, it will support high data rate services, including 2.048 Mbit/s for indoor users, 384 Kbit/s for pedestrian subscribers, 144 Kbit/s for moving vehicles, 9.6 Kbit/s for mobile satellite services. The radio technologies are expected to have vastly improved capabilities over existing 2G mobile systems (e.g., multi-environment, multi-mode, multi-band, multimedia operations).

8.9.2 CDMA Proposals

A number of Radio Transmission Technologies (RTTs) developed in Asia, Europe, and the US have been proposed for IMT-2000. Three proposals were initially submitted, namely direct-sequence wideband CDMA (W-CDMA), multi-carrier CDMA mode (CDMA2000), and time-duplexed CDMA (TD-CDMA). W-CDMA is backward compatible to GSM and PDC systems in Europe and Asia respectively while CDMA2000 is backward compatible to IS-95 (developed by the TIA) in North America. Recently, the wireless industry has agreed to a harmonized Global 3G (G3G) CDMA framework.

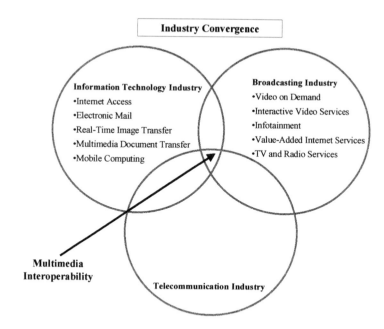

Figure 8.16: Convergence of disparate technologies in IMT-2000

8.9.3 Wideband CDMA

W-CDMA increases data rates by using multiple 1.25 MHz channels as opposed to the single channel adopted by the current IS-95 standard (Figure 8.17). The main W-CDMA specifications are listed in Table 8.8. TD-CDMA on the other hand offers backward compatibility with the huge network infrastructure investment operators have made in standards such as GSM.

A significant concept distinguishing W-CDMA from current IS-95 systems is the introduction of inter-cell asynchronous operation, which is vital for continuous system deployment from outdoors to indoors, and the data channel associated pilot channel for a coherent reverse link as well as a forward link. W-CDMA facilitates the application of interference cancellation and adaptive antenna array techniques on both the reverse and forward links to significantly enhance the link capacity and coverage.

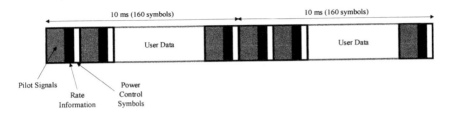

Figure 8.17: W-CDMA uplink multirate transmission

Table 8.8: W-CDMA parameters

Parameter	Specification
Channel bandwidth (MHz)	1.25, 5, 10 and 20
Chip rate (Mchips/s)	1.024, 4.096, 8.192, 16.384
Frame length (ms)	10, 20
Data modulation	QPSK (downlink), BPSK (uplink)
Multirate	Variable spreading and multicode
Power control	Open and fast closed loop control
Spread factors	4 to 256

8.9.4 Time-Duplexed CDMA

In TD-CDMA, different channels are multiplexed onto the same time slot. Since the spreading ratio is small, it may require multiuser detection to remove intracell interference. Another reason why multiuser detection is needed can be attributed to the slow power control in TD-CDMA, resulting in highly variable received power levels. The main TD-CDMA specifications are listed in Table 8.9.

8.9.5 CDMA2000

The main goal of CDMA2000 is to provide 144 Kbit/s and 384 Kbit/s with approximately 5 MHz of bandwidth. Like W-CDMA, CDMA2000 employs slotted ALOHA for packet transmission. However, instead of a fixed transmit power, it increases the transmit power after an unsuccessful attempt. This increases the probability of success for a retransmission through the capture effect. The parameters for this standard are listed in Table 8.10.

Table 8.9: TD-CDMA parameters

Parameter	Specification
Channel bandwidth (MHz)	1.6
Chip rate (Mchips/s)	2.167
Frame length (ms)	4.615 (8 slots per frame)
Data modulation	QPSK, 16-QAM
Multirate	Multislot and multicode
Signal detection	Coherent
Spread factors	16 chips/symbol

Table 8.10: CDMA2000 parameters

Parameter	Specification
Channel bandwidth (MHz)	1.25, 5, 10, 15 and 20
Direct-spread chip rate (Mchips/s)	1.2288, 3.6864, 7.3728, 11.0593, 14.7456
Multicarrier chip rate (Mchips/s)	N x 1.2288 (N = 1, 3, 6, 9, 12)
Frame length (ms)	5 (control), 20 (data)
Data modulation	QPSK (downlink), BPSK (uplink)
Multirate	Variable spreading and multicode
Power control	Open loop and fast closed loop controls
Spread factors	4 to 256

8.9.6 TDMA Proposals

TDMA technology is represented by Universal Wireless Communication (UWC-136), which harmonizes GSM with North America's TDMA. The CDMA and TDMA air-interface proposals were harmonized through the 3G-Partnership Project (3GPP).

A second world standard is expected as GSM and TDMA systems evolve into the 3G standard, Enhanced Data rates for GSM Evolution (EDGE). EDGE increases the data rate of current GSM systems by roughly three times through the use of 8-PSK (3 bits/symbol) modulation instead of GMSK (1 bit/symbol). The basic parameters of the EDGE packet radio standard are listed in Table 8.11.

Table 8.11: Parameters for GSM and EDGE

Radio Parameters	GSM	EDGE
Carrier spacing	200 KHz	200 KHz
Modulation rate	270.1 symbols/s	270.1 symbols/s
Frame length	4.615 ms	4.615 ms
Number of slots/frame	8	8
Modulation	GMSK	8-PSK
Payload per burst (symbols)	116	116
Bits per burst (bits)	116	384
Radio interface data rate	22.8 Kbit/s (1 slot)	69.6 Kbit/s (1 slot)
	182.4 Kbit/s (1 frame)	556.8 Kbit/s (1 frame)

SUMMARY

Many third-generation wireless systems involving high-speed wireless LANs, wireless ATM networks, and wireless Internet connectivity are the major focus of recent research efforts. These broadband networks aim to provide integrated, packet-oriented, transmission of text, graphics, voice, image, video, and computer data between individuals as well as in the broadcast mode. Although the underlying access protocols supporting these networks have evolved rapidly, the basic access methods (e.g., ALOHA, CSMA, TDMA, CDMA) are still very much relevant.

BIBLIOGRAPHY

[ANAS98]   Anastasi, G., Lenzini, L., Mingozzi, E., Hettich, A. and Kramling, A., "MAC Protocols for Wideband Wireless Local Access: Evolution Toward Wireless ATM", *IEEE Personal Communications*, October 1998, pp. 53 – 64.

[BANT94]   Bantz, D. and Bauchot, F., "Wireless LAN Design Alternatives", *IEEE Network*, Vol. 8, No.2, March/April 1994, pp. 43 – 53.

[BEND00]   Bender, P., Black, P., Grob, M., Padovani, R., Sindhushayana, N. and Viterbi, A., "CDMA/HDR: A Bandwidth Efficient High Speed Wireless Data Service for Nomadic Users", *IEEE Communications Magazine*, July 2000.

[BISW92]   Biswas, S., Porter, J. and Hopper, J., "Performance of a Multiple Access Protocol for an ATM-Based Picocellular Radio LAN", *Proceedings of the 3rd IEEE Conference on Personal, Indoor, and Mobile Radio Communications*, 1992, pp. 139 – 144.

[BING00]   Bing, B., *High-Speed Wireless ATM and LANs*, Artech House, 2000.

[BORG96]   Borgonovo, F., Fratta, L., Zorzi, M. and Acampora, A., "Capture Division Packet Access: A New Cellular Access Architecture for Future PCNs", *IEEE Communications Magazine*, September 1996, pp. 154 – 162.

[CHEN94]   Chen, K., "Medium Access Control of Wireless LANs for Mobile Computing", *IEEE Network Magazine*, Vol. 8, No. 5, September 1994, pp. 50 – 63.

[CHHA96]   H. Chhaya and Gupta, S., "Performance of Asynchronous Data Transfer Methods of IEEE 802.11 MAC Protocol", *IEEE Personal Communications*, Vol. 3, No. 5, October 1996, pp. 8 – 15.

[CHUA00]   Chuang, J. and Sollenberger, N., "Beyond 3G: Wideband Wireless Data Access based on OFDM and Dynamic Packet Assignment", *IEEE Communications Magazine*, July 2000.

[DELL97]   Dellaverson, L. and Dellaverson, W., "Distributed Channel Access on Wireless ATM Links", *IEEE Communications Magazine*, Vol. 35, No. 11, November 1997, pp. 110 – 113.

[GARG00]   Garg, V., *IS-95 CDMA and CDMA 2000: Cellular/PCS Systems*, Prentice Hall, 2000.

[HAAR00]   Haartsen, J., "The Bluetooth Radio System", *IEEE Personal Communications*, Vol. 7, No. 1, February 2000, pp. 28 – 36.

[HUNG98]   Hung, A., Montepit, M. and Kesidis, G., "ATM via Satellite: A Framework and Implementation", *ACM/Baltzer Journal on Wireless Networks*, Vol. 4, 1998, pp. 141 – 153.

[IEEE96]   "Special Issue on Wireless ATM", *IEEE Personal Communications*, Vol. 3, No. 4, August 1996.

[IEEE97]   IEEE Std 802.11, *Information Technology – Telecommunications and Information Exchange Between Systems – Local and Metropolitan Area Networks – Specific Requirements, Part 11: Wireless LAN Medium Access Control (MAC) and Physical Layer (PHY) Specifications*, November 1997.

[IEEE99a] IEEE Std 802.11a, *Supplement to Standard for Information Technology – Telecommunications and Information Exchange Between Systems – Local and Metropolitan Area Networks – Specific Requirements, Part 11: Wireless LAN Medium Access Control (MAC) and Physical Layer (PHY) Specifications: High Speed Physical Layer in the 5 GHz Band*, September 1999.

[IEEE99b] IEEE Std 802.11b, *Supplement to Standard for Information Technology – Telecommunications and* Information Exchange Between *Systems – Local and Metropolitan Area Networks – Specific Requirements, Part 11: Wireless LAN Medium Access Control (MAC) and Physical Layer (PHY) Specifications: Higher Speed Physical Layer Extension in the 2.4 GHz Band*, September 1999.

[IEEE99c] "Special Issue on the Evolution of TDMA to 3G", *IEEE Personal Communications Magazine*, Vol. 6, No. 5, June 1999.

[KARO95a] Karol, M., Liu, Z. and Eng, K., "An Efficient Demand-Assignment Multiple Access Protocol for Wireless Packet (ATM) Networks", *ACM/Baltzer Journal on Wireless Networks*, Vol. 1, pp. 267 – 279.

[KARO95b] Karol, M., Liu, Z. and Eng, K., "Distributed-Queueing Request Update Multiple Access (DQRUMA) for Wireless Packet (ATM) Networks", *Proceedings of the IEEE ICC*, 1995, pp. 1224 – 1231.

[KARO95c] Karol, M., Haas, Z. and Woodsworth, C., "Performance Advantages of Time-Frequency Sliced Systems", *Proceedings of 6th IEEE PIMRC*, 1995, pp. 1104 – 1109.

[KARO96] Karol, M., Liu, Z. and Pancha, P., "Implications of Physical Layer Overhead on the Design of Multiaccess Protocols", *Proceedings of the IEEE ICC*, 1995, pp. 1224 – 1231.

[KIM00] Kim, K. (editor), *Handbook of CDMA Systems Design, Engineering and Optimization*, Prentice Hall, 2000.

[LUIS98] Luise, M. and Pupolin, S., *Broadband Wireless Communications: Transmission, Access, and Services*, Springer-Verlag, 1998.

[MIKK98] Mikkonen, J., Aldis, J., Awater, G., Lunn, A. and Hutchison, D., "The Magic Wand – Functional Overview", *IEEE Journal on Selected Areas in Communications*, Vol. 16, No. 6, August 1998, pp. 953 – 972.

[NEGU00] Negus, K., Stephens, A., Lansford, J., "HomeRF: Wireless Networking for the Connected Home", *IEEE Personal Communications*, Vol. 7, No. 1, February 2000, pp. 20 – 27.

[OONO97] Oono, T., Takanashi, H. and Tanaka T., "Dynamic Slot Allocation Technologies for Mobile Multimedia TDMA Systems using a Distributed Scheme", *6th IEEE International Conference on Universal Personal Communications*, October 1997.

[RAYC99] Raychaudhuri, D., "Wireless ATM Networks: Technology Status and Future Directions", *Proceedings of the IEEE*, Vol. 87, No. 10, October 1999, pp. 1790 – 1806.

[SANC97] Sanchez, J., Martinez, R. and Marcellin, M., "A Survey of MAC Protocols for Wireless ATM", *IEEE Network*, Vol. 11, No. 6, November/December 1997, pp. 52 – 62.

[STEE99] Steele, R. and Hanzo, L. (editors), *Mobile Radio Communications: Second and Third-Generation Cellular and WATM Systems*, John Wiley, 1999.

[UMEH96] Umehira, M., Nakura, M., Sato, H. and Hashimoto, A., "ATM Wireless Access for Mobile Multimedia: Concept and Architecture", *IEEE Personal Communications*, October 1996, pp. 39 – 48.

[WANG98] Wang, M. and Kohno, R., "A Wireless Multimedia DS-CDMA Network based on Adaptive Transmission Rate/Power Control", *1998 International Zurich Seminar on Broadband Communications*, pp. 45 – 49.

[ZHAN94] Zhang, C., Hafez, H. and Falconer, D., "Traffic Handling Capacity of a Broadband Indoor Wireless Networks Using CDMA Multiple Access", *IEEE Journal on Selected Areas in Communications*, Vol. 12, No. 4, May 1994, pp. 645 – 652.

Chapter 9

A GENERALIZED BROADBAND WIRELESS
ACCESS PROTOCOL

A generalized access protocol that fulfils the demanding requirements dictated by multimedia applications operating in broadband wireless networks is described and analyzed. Instead of maintaining a fixed number of reservation slots in a periodic manner, the number of reservation slots in this protocol is a function of the current estimate of the traffic load of the network. An analytical approach based on queuing theory is employed to gain some insight into the behavior of the protocol. This approach appears to be accurate in predicting its performance for both bursty and periodic traffic types. The analytical results are verified by network simulation. The protocol is modeled using OPNET. A formal verification for part of the protocol is also provided. The unique ability of the protocol in improving the bandwidth sharing among different cells in a wireless network is also demonstrated.

9.1 PROTOCOL DESCRIPTION

As explained in Section 2.2.4, bandwidth must be guaranteed for an application not only to satisfy the bandwidth requirement but also to limit the delay and errors introduced. For a broadband application, the QoS requirements must also be maintained. Hence, a broadband multiple access protocol should be designed in such a way that each new application first negotiate with the network for available bandwidth resource before a connection is established. Second, the protocol should guarantee the bandwidth for the application once it is accepted by the network. This implies two criteria for the protocol:

❑   connection-oriented;
❑   reservation-based.

The generalized scheme is in essence, a contention-based reservation protocol. The method described here assumes centralized control. Base stations accept bursty traffic and explicit reservations and then schedule the times for users with periodic traffic to send or receive data. Bursty traffic specifically refers to short-duration interactive or priority messages (e.g., user registrations, handoff requests) called bursty frames. Periodic traffic is mostly VBR traffic involving compressed voice, video. Such traffic requires reservations to transmit a long message that is segmented into one or more periodic frames. Typically, the size of a periodic frame is several times greater than that of a bursty frame. The overall channel transmission time is divided into two distinct cycles:

❑ A Contention Cycle for the transmission of bursty traffic and reservations on a random access basis;
❑ A Message Cycle for the collision-free transmission of periodic traffic.

The protocol operates with Contention Cycles until one or more successful reservations are received, after which the protocol reverts between Contention and Message Cycles. Figure 9.1 depicts the possible events related to the operation of the proposed protocol. Dark regions represent transmissions that originate from the base station. Flowcharts depicting the operation of the protocol are shown in Figures 9.2 to 9.4. For simplicity, it is assumed that the base station does not send messages to the users and that reservations are processed on a FCFS basis by the base station.

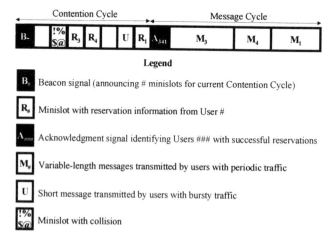

Figure 9.1: Operation of the generalized protocol

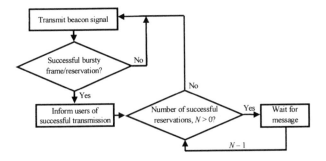

Figure 9.2: Operation of the base station

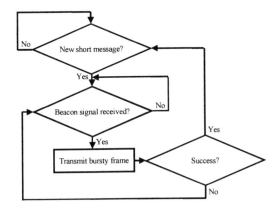

Figure 9.3: Operation of the user with bursty traffic

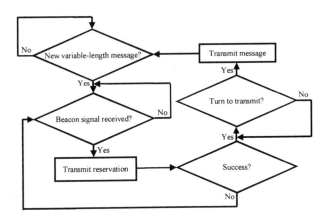

Figure 9.4: Operation of the user with periodic traffic

The base station initiates a new Contention Cycle by announcing the number of minislots in the cycle through the beacon signal. The beacon signals must be transmitted before the users are able to transmit their requests. The length and number of the minislots are variable parameters assigned by the base station depending on the proportion of bursty and periodic traffic in the network. Typically, the number of minislots grows larger as the traffic load increases. The number of minislots declared by the base station allows current users operating in a radio cell as well as potential users from adjacent cells to estimate the traffic loading (Section 9.6). An optional feature enables slots to be partitioned into different categories such as real-time traffic or non-realtime traffic. In this way, priority can be incorporated by assigning a different number of slots for different types of service requests.

Conventional slotted ALOHA (where users with new arrivals transmit immediately on the next slot) is not an appropriate protocol to be used for contention since a high collision probability is likely on the first minislot. This is particularly true after the completion of a long Message Cycle which increases the probability of new messages and bursty frames accumulating during the cycle. Thus, contention efficiency can be improved by ensuring that all users randomly select a minislot from the Contention Cycle. For the generalized protocol, a ready user sends a bursty frame or reservation in one of the minislots with a probability of $1/V$ where V represents the total number of minislots in a Contention Cycle. This technique is called the randomized slotted ALOHA (RSA) protocol. From simulation studies, a value of $V \geq 6$ appears to be a reasonable choice for a wide range of traffic loads and user densities while a value of $V \geq 3$ is suitable for low to medium traffic loads.

At the end of the Contention Cycle, users with successfully transmitted bursty frames/reservations will be informed simultaneously by the base station through the acknowledgment signal. The acknowledgment signal also informs users the order and transmission time they are allocated. A ready user intending to send a variable-length message or establish a connection will first need to succeed in sending a reservation in any of the minislots. The reservation contains information defining the identity of the user, the message length, the message priority, the traffic repetition interval and the desired transit delay (holding time) relative to this interval. An ongoing connection can always be maintained by piggybacking new reservations on currently transmitted messages. This eliminates the need to make regular reservations when the future traffic requirements of a user change, thereby reducing congestion in the Contention Cycle. To prevent such users from monopolizing channel transmission time, the number of periodic frames is limited to a maximum for each Message Cycle. Similarly, a user with a short message to send will transmit a bursty frame in any of the minislots. Although a bursty frame may request information from the base station, it is unable to negotiate for channel transmission time.

Unsuccessful users attempt retransmission in successive Contention Cycles until the bursty frame/reservation is successfully transmitted. If the number of attempts exceeds a pre-determined threshold, the request is discarded. The first message or bursty frame that arrives after the beacon signal will have to wait for the next beacon signal. In this way, idle users do not have to monitor the beacon signal continuously and such users can conserve power by operating in the low power (sleep) mode. This function is particularly important for battery-powered terminals.

All successful reservations are entered into a global (reserved) queue that is maintained by the base station. These requests are ordered according to reservation urgency which is a function of priority and residual lifetime. Whilst a lower priority reservation is waiting, its residual lifetime will be decremented. The base station may decide to increase the priority of a reservation as its residual lifetime decreases. When the residual lifetime becomes zero and the reservation has not been serviced, it will be discarded. Within the same priority class, FCFS policy is adopted.

9.2 PROTOCOL EVLUATION

The centralized nature of protocol provides an ideal topology for both portable and mobile users. Such users can listen to the beacon signal from each base station within range. During this listening period, the base station with the best reception is selected. This can be done by transmitting a backward handoff request (Section 1.9.3) via the Contention Cycle to the current base station to request for the connection to be transferred to the target base station.

In the generalized scheme, short Contention Cycles dominate when the network loading is light and bursty frames/reservations get transmitted across the network quickly. Under heavy traffic conditions, the channel bandwidth is efficiently utilized when users with periodic traffic transmit their messages. Thus, the proposed protocol adapts automatically to a time-varying traffic load.

The generalized protocol allows users that generate VBR traffic to send messages of arbitrary duration. Variable message lengths are more efficient than time-slotted transmission for two main reasons. First, if slots are chosen to match the largest lengths, slot times that are under-utilized by short messages must be padded out to fill up the slots. On the other hand, if smaller slot sizes are used, more overhead per message results. Second, the total number of packets sent is minimized. This can be important in obtaining high throughput since many network devices are limited not by the number of bits they can transmit per second but by the number of packets they can process per second.

Delay variance (jitter) due to variable message lengths can be mitigated by employing additional buffers to restore synchronization. Moreover, since information about the statistical needs of all ready users can be gathered through reservations, service order and duration can be dynamically controlled to attain certain QoS without compromising optimum system performance.

9.3 ANALYSIS OF RANDOMIZED SLOTTED ALOHA

In this section, the RSA protocol is analyzed. The operation of the protocol is shown in Figure 9.5. At the start of each frame, a beacon signal defines the total number of minislots (V) in the current frame and acknowledges the successful transmissions in the slots of the previous frame. The value for V changes depending on the proportion of the channel load estimated by the base station. A ready user chooses one of the V slots at random with an equal probability of $p = 1/V$ (i.e., $\sum_{i=1}^{V} p_i = 1$). A ready user can either be in the originating mode or in the backlogged (retransmission) mode. A transmission is successful only when one user attempts transmission at a minislot. When two or more users transmit at a minislot, their transmissions collide and none will be successful. A collided packet must be retransmitted in subsequent Contention Cycles until it succeeds.

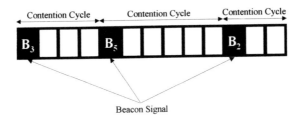

Beacon Signal

Figure 9.5: Operation of the randomized slotted ALOHA Protocol

9.3.1 Throughput Analysis

Denote:

E = Average number of retransmissions for each successful transmission.
G = Average number of attempted transmissions for each minislot.
S = Average number of successful transmissions for each minislot.

An approximate throughput expression can be derived by assuming that all attempted transmissions from each of the V slots form an aggregate Poisson process with a total arrival rate of VG. For a network with infinite users, the probability of exactly one transmission in the current slot is given by:

$$S = \sum_{i=1}^{\infty} \frac{(VG)^i e^{-VG}}{i!} \binom{i}{1}\left(\frac{1}{V}\right)\left(1-\frac{1}{V}\right)^{i-1}$$

$$= \frac{VGe^{-VG}}{V} \sum_{i=1}^{\infty} \frac{(VG-G)^{i-1}}{(i-1)!} = \frac{VGe^{-VG}}{V} \times e^{VG-G} = Ge^{-G}$$

$$(9.1)$$

which is identical to the throughput for conventional slotted ALOHA (equation 5.7). Alternatively, this same throughput expression can be obtained by considering the probability of a successful transmission from arrivals that are generated in one of the minislots and then summing all probabilities over V slots.

$$S = \binom{V}{1} \left[\sum_{i=1}^{\infty} \frac{G^i e^{-G}}{i!} \binom{i}{1} \left(\frac{1}{V} \right) \left(1 - \frac{1}{V} \right)^{i-1} \right] \left[\sum_{i=0}^{\infty} \frac{G^i e^{-G}}{i!} \left(1 - \frac{1}{V} \right)^{i-1} \right]^{V-1}$$

$$= V \left(\frac{G e^{-G/V}}{V} \right) \left(e^{-G/V} \right)^{V-1} = G e^{-G}$$

(9.2)

In equilibrium, the ratio G/S equals the average number of transmissions for each successful transmission. Thus, the following relationship holds:

$$E = \frac{G}{S} - 1 = e^G - 1$$

(9.3)

The key assumption in the above analysis is that the offered channel load, G, which includes retransmissions, forms a Poisson process. This assumption can be a good approximation at best. Under heavy network loading, due to a much higher retransmission rate, the Poisson assumption becomes invalid. Studies have shown, however, that if the retransmission schedule is chosen uniformly from an arbitrarily large interval, then the number of scheduling points in any interval approaches a Poisson distribution.

9.3.2 Stability Analysis

It is well known that random access protocols operating with finite Poisson sources are intrinsically bistable [CARL75], [FAYO77], [HANS79] and must be equipped with proper controls. Otherwise, statistical fluctuations in the channel load may cause the system to drift to a saturated state where collisions prevail and low throughput results.

To avoid such undesirable behavior for the RSA protocol, V must somehow adapt to the number of ready users (or equivalently, the channel load). Specifically, V should increase when the number of ready users is large (thereby reducing p). On the other hand, when the number of ready users is small, V should decrease so that the average access delay is minimized under conditions of low channel load. Thus, to control V (and hence p) dynamically, an estimation of the number of ready users is required. The main problem here is that the number of ready users is unknown (to the controller) and varying.

Recognizing this problem, many dynamic control policies have been suggested for random access protocols and they generally require binary or ternary feedback information to estimate the number of ready users [BERT92], [KELL85], [RIVE87], [ROM90]. However, a multiple access protocol is more robust (in terms of insensitivity to noise or other degradations) when it does not depend on feedback information about past transmissions makes.

To investigate whether stability can be maintained for the RSA protocol even when the estimate for V is inaccurate, consider a finite population of N ready users accessing a central controller using the RSA protocol. Each user generates fixed-length packets independently. These assumptions are made identical to those in [CARL75], [HANS79], [JENQ81] in order to facilitate comparison between conventional slotted ALOHA and the RSA protocol analyzed here. The overhead due to the beacon signal is neglected.

Let $r(t)$, $t = 1, 2, 3, \dots$ be the number of backlogged users at the beginning of slot t. The quantity $r(t)$ can assume one of $(N + 1)$ possible values of $\{0, 1, 2, \dots, N\}$ and can be considered as the state variable of the system. Since the system is memoryless, $r(t)$ is a finite Markov chain with the transition probability matrix $P = [p_{n,m}]$ where $p_{n,m} = \text{Prob}[r(t + 1) = m \mid r(t) = n]$ is given by:

$$
p_{n,m} = \begin{cases}
0, & m \leq n - 2 \\
np(1 - p)^{N-1}, & m = n - 1 \\
(1 - p)^{N-n} + (N - 2n)p(1 - p)^{N-1}, & m = n \\
(N - n)p(1 - p)^{N-n-1}\left[1 - (1 - p)^n\right], & m = n + 1 \\
\binom{N - n}{m - n} p^{m-n}(1 - p)^{N-m}, & m \geq n - 2
\end{cases}
$$

$$(9.4)$$

Note that for any probability $0 < p < 1$, this Markov chain is finite, irreducible and aperiodic. Under these conditions, the Markov chain is always stable in the sense that with time, $\text{Prob}[r(t) = m]$, $m = 0, 1, ..., N$ converges to a valid probability mass function (i.e., all the probabilities are non-negative and they add to 1).

The average throughput per slot, $S(n, p, N)$, when the system is in state n is:

$$S(n, p, N) = p_{n,n-1} + (N - n)p(1 - p)^{N-1} = Np(1 - p)^{N-1}$$

(9.5)

For large N:

$$S(n, p, N)_{max} = S(n, 1/N, N) = (1 - 1/N)^{N-1} \approx 1/e$$

(9.6)

This implies that if N is known to the base station, then the throughput can be maximized by ensuring that each ready user transmits at a time slot with a probability equal to $1/N$. It is not practical to obtain N in a mobile wireless network since N varies dynamically as users enter and depart from the network.

The average drift or the expected change in the number of backlogged users is a useful parameter that facilitates stability studies of random access protocols [CARL75], [HANS79], [JENQ81]. The average drift in state n is obtained by subtracting the channel output rate (i.e. the throughput of the system) from the channel input rate, $(N - n)p$. A positive drift means that the number of backlogged users tend to grow larger while a negative drift has the opposite effect. When the average drift is zero, the input and output packet flows are balanced and therefore, the system is in equilibrium.

Equilibrium points with positive-to-negative going average drift are stable [CARL75], [JENQ81]. This can be explained as follows. Fluctuations in the number of backlogged users below the equilibrium point tend to increase since the average drift is positive. Conversely, fluctuations above the equilibrium point tend to decrease since the average drift is negative.

For the RSA protocol, the average drift, $D(n, p, N)$, in state n is:

$$D(n, p, N) = (N - n)p - Np(1 - p)^{N-1}$$

(9.7)

Rewriting $D(n, p, N)$ with $n = xN$ and $p = 1/V = 1/(yN)$:

$$D\left(xN, \frac{1}{yN}, N\right) = \frac{1}{y} - \frac{x}{y} - \frac{1}{y}\left(1 - \frac{1}{yN}\right)^{N-1} \qquad \text{where } 0 \leq x \leq 1 \text{ and } yN \geq 1$$

(9.8)

Consider the first case when the estimate for N is exact (i.e., when $y = 1$). A plot of the average drift for various values of x and N is shown in Figure 9.6. Clearly, the system is stable at one equilibrium point and is relatively independent of N for large N. Furthermore, when N is large, equation 9.4 (and Figure 9.6) shows that n approaches $0.63N$ at the equilibrium point. Although this implies that an appreciable number of arriving packets are discarded and that the delay is large, it is important to emphasize that for the RSA protocol, new arrivals are held up only when the system is already estimated to be congested. When the estimated channel load is low (i.e., when yN is low), p can be as large as 1.

Consider now the case when the estimate for N is inaccurate. By varying x, y, and N, Figure 9.7 is obtained. Once again, the system possesses a single equilibrium point. An underestimated N (i.e., when $0 < y < 1$) increases the number of backlogged users at the equilibrium point. The rate of increase is faster compared to the case when N is overestimated (i.e., when $y > 1$). For example, if $N = 20$ and $y = 0.5$ (50% under-estimation), 0.9 (10% under-estimation), 1.0 (exact), 1.1 (10% over-estimation), 1.5 (50% over-estimation), then the system is in equilibrium at $x = 0.865, 0.662, 0.623, 0.587, 0.475$ respectively. The rate of deviation in x from the exact case is obvious when N becomes increasingly underestimated.

As depicted in Figure 9.8, the average throughput also deteriorates quickly. Under the worst case when $yN = 1$, the system is fully backlogged at the equilibrium point and the corresponding throughput reduces to zero. Intuitively, this is expected since when there is one slot in a frame (i.e., when $V = yN = 1$), a single collision will trigger an indefinite number of retransmissions, thus leading to zero throughput. This situation can be avoided through the use of collision feedback or by simply imposing a minimum of two or more slots for V. The latter method suggests that the RSA protocol can be made stable without the use of channel feedback. This property differs significantly from conventional slotted ALOHA which achieves stable throughput by employing channel feedback.

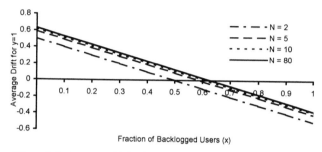

Figure 9.6: Average drift when channel load estimate is exact

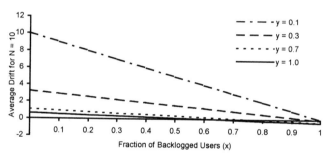

Figure 9.7: Average drift when channel load estimate varies

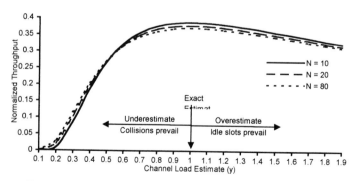

Figure 9.8: Throughput as a function of channel load estimate

9.4 ANALYSIS OF THE GENERALIZED PROTOCOL

An analytical model under conditions of steady-state equilibrium is studied. A list of symbols related to the analysis is provided in Table 9.1.

The general assumptions are:

- ❑ The network consists of a reasonably large number of users that are fully-connected to one base station;
- ❑ Each user has an independent Poisson arrival and holds at most one bursty frame or message;
- ❑ The length of a bursty frame and a reservation are each equivalent to the duration of one minislot;

Table 9.1: List of symbols

Symbol	Description
C	Overall channel capacity (in periodic frames per second)
E	Average number of retransmissions (bursty frames and reservations)
F	Fraction of bursty frames generated
G	Normalized offered load for each minislot
G_{sys}	Normalized overall offered load
M	Average message length (in periodic frames)
P	Average Message Cycle length (in periodic frames)
P_p	Probability of Message Cycle occurrence
R	Average Contention Cycle length (in periodic frames)
S_b	Normalized throughput for bursty frames
S_r	Normalized throughput for reservations
V	Total number of minislots in a Contention Cycle
λ	Total number of new arrivals averaged over one second
ρ_b	Normalized channel utilization for bursty frames
ρ_m	Normalized channel utilization for messages
E_b	Average number of retransmitted bursty frames
E_p	Average number of retransmitted reservations
$\overline{M^2}$	Second moment of message length
$\overline{R^2}$	Second moment of Contention Cycle
T_b	Average delay for bursty traffic
T_p	Average delay for periodic traffic
W_b	Average waiting time for bursty frames
W_p	Average waiting time for reservations
Y_b	Average bursty frame transmission time
Y_p	Average message transmission time

- All message lengths are independent and identically distributed random variables;
- Beacon and acknowledgment signals do not incur overheads;
- The round-trip channel propagation delay is neglected.

9.4.1 Throughput Analysis

In the analysis, channel utilization is defined as the fraction of the total channel capacity that is dedicated to the successful transmission of bursty frames or messages. Throughput refers to the normalized rate at which bursty frames or messages are successfully transmitted. Equations 9.9 to 9.14 can be derived easily from first principles.

$$S_b = \frac{F\lambda \times \dfrac{R}{V}}{\dfrac{R}{P+R} \times C} = \frac{F\lambda(P+R)}{CV}$$

(9.9)

$$S_r = \frac{(1-F)\lambda \times \dfrac{R}{V}}{\dfrac{R}{P+R} \times C} = \frac{(1-F)\lambda(P+R)}{CV}$$

(9.10)

$$S_m = \frac{(1-F)\lambda M}{\dfrac{P}{P+R} \times C} = \frac{(1-F)\lambda M(P+R)}{CP}$$

(9.11)

$$\rho_b = \frac{F\lambda R}{CV} = \frac{S_b R}{P+R}$$

(9.12)

$$\rho_r = \frac{(1-F)\lambda R}{CV} = \frac{S_r R}{P+R}$$

(9.13)

$$\rho_m = \frac{(1-F)\lambda M}{C} = \frac{S_m P}{P+R}$$

(9.14)

The utizations ρ_m, ρ_r, and ρ_b can be obtained directly by combining a Contention Cycle and a Message Cycle to form a CYCLE. In the steady state, all CYCLEs are statistically similar. Thus, the channel utilization can be determined as the ratio of the average amount of time during a CYCLE devoted to successful transmissions to the average total length of the CYCLE.

Since $S_m = 1$, from equation 9.11,

$$P = \frac{\rho_m R}{1-\rho_m}$$

(9.15)

Using equations 9.12 and 9.14,

$$\rho_b = \frac{FR\rho_m}{(1-F)MV}$$

(9.16)

From equations 9.9, 9.10, and 9.15,

$$S_b + S_r = \frac{P}{(1-F)MV} \approx Ge^{-G}$$

(9.17)

Now

$$G_{sys} \approx \frac{\frac{\lambda R}{V} + (1-F)\lambda M + E\frac{\lambda R}{V}}{C} = (1-\rho_m)G + \rho_m$$

(9.18)

Intuitively, equation 9.18 makes sense since the system is busy for the fraction ρ_m of the time.

From equation 9.17,

$$0 \leq P \leq \frac{(1-F)MV}{e}$$

(9.19)

Using equations 9.15 and 9.19,

$$0 \leq \rho_m \leq \frac{(1-F)MV}{(1-F)MV + eR}$$

(9.20)

It is interesting to observe that when $G = 1$, $G_{sys} \approx 1$. This means that the channel utilization for both bursty and periodic traffic (and hence the overall channel utilization) is maximized when the normalized value of the total offered traffic is close to unity.

9.3.2 Delay Analysis

The average delay is defined as the average time a new bursty frame or message arrives at the user terminal's buffer to the time the complete frame or message is correctly received by the base station. This delay is normalized to periodic frames. Figures 9.9 and 9.10 show the various components required in the calculations. In all cases, the average delay is computed in units of periodic frames. Note that $E_b = FE$ and $E_p = (1 - F)E$.

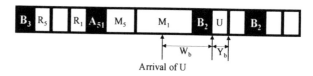

Figure 9.9: Average delay for bursty traffic

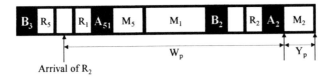

$$\text{Figure 9.10: Average delay for periodic traffic}$$

A bursty frame will have to wait for a residual time until the end of the current message transmission or Contention Cycle. It must also wait for the transmission of messages due to reservations made before its successful transmission (queuing delay). If an attempted transmission results in a collision, the collided frame will have to be retransmitted and each retransmission incurs an additional delay of R periodic frames. The bursty frame is successfully transmitted after $(V-1)R/2V$ periodic frames on the average.

The average waiting time for periodic traffic is similar to that for bursty traffic except that a delay of R periodic frames is incurred before a message is successfully transmitted.

For $0 \le \rho_m < 1$,

$$W_b = \left(1 - \rho_m\right)\frac{\overline{R^2}}{2R} + \rho_m \frac{\overline{M^2}}{2M} + \rho_m W_b + E_b R + \frac{(V-1)R}{2V}$$

$$(9.21)$$

$$W_p = \left(1 - \rho_m\right)\frac{\overline{R^2}}{2R} + \rho_m \frac{\overline{M^2}}{2M} + \rho_m W_p + E_p R + R$$

$$(9.22)$$

Since $Y_b = R/V$,

$$T_b = Y_b + W_b = \frac{R}{V} + \frac{\overline{R^2}}{2R} + \frac{\rho_m \dfrac{\overline{M^2}}{2M} + E_b R + \dfrac{(V-1)R}{2V}}{\left(1 - \rho_m\right)}$$

$$(9.23)$$

Since $Y_p = M$,

$$T_p = Y_p + W_p = M + \frac{\overline{R^2}}{2R} + \frac{\rho_m \frac{\overline{M^2}}{2M} + (E_p + 1)R}{(1 - \rho_m)}$$

$$(9.24)$$

The operation of the generalized protocol is simulated using OPNET (Section 9.7). Extensive simulation carried out using different system parameters shows good agreement with the results obtained from the analytical model. The theoretical limits of P and ρ_m given by equations 9.19 and 9.20 respectively have also been validated. The network parameters are:

❑ A population of 40 users generating traffic at equal arrival rates;
❑ Each Contention Cycle has a constant length of 1 periodic frame;
❑ The message length is either fixed at 1 periodic frame (i.e. $M = 1$) or allowed to follow a discrete uniform distribution that varies from 1 to 8 periodic frames (i.e., $M = 4.5$);
❑ Only bursty frames and messages contribute to the effective (useful) channel utilization while reservations are considered as overheads.

A sample of the numerical results from both analysis and simulation is presented here (Figures 9.11 to 9.17) to illustrate the performance of the proposed protocol and to assess the accuracy of the analytical results. In the figures, F denotes the fraction of bursty frames generated while M is the average message length (in periodic frames). Although V is chosen to be fairly large in the simulation, there is still a slight discrepancy in some of the results at high load. This can be attributed to the approximation that the offered channel load G forms a Poisson process (Section 9.3.1).

Equation 9.18 suggests that the channel utilization is maximized when the normalized value of the overall offered load is close to unity. This has been verified in Figure 9.11. Figures 9.13 to 9.15 show that the length of the Message Cycle directly affects the average delay performance. Note that under low traffic load, the average delay for bursty traffic is hardly affected by the presence of periodic traffic (compare Figure 9.12 with Figures 9.14 and 9.15). This implies that bursty traffic can still be transmitted quickly under such traffic conditions. Figure 9.17 suggests that there will be more Message Cycles if the average message length (M) is short. However, a reduction in the overall channel utilization results.

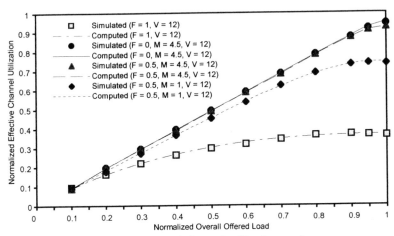

Figure 9.11: Normalized channel utilization performance

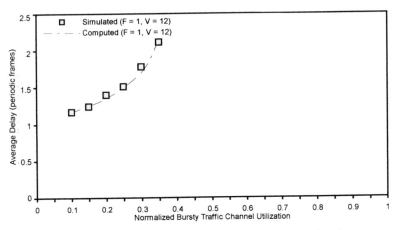

Figure 9.12: Average delay performance $(F = 1, M = 0)$

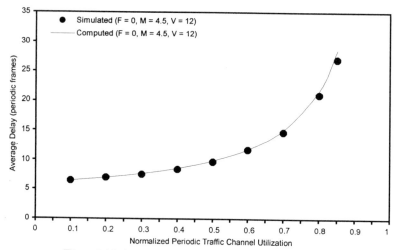

Figure 9.13: Average delay performance $(F = 0, M = 4.5)$

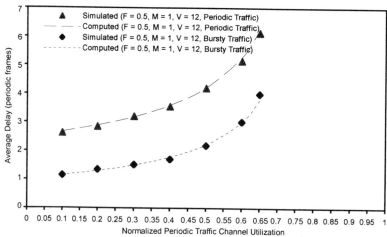

Figure 9.14: Average delay performance $(F = 0.5, M = 1)$

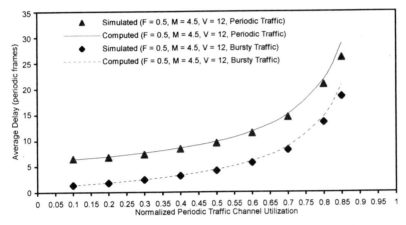

Figure 9.15: Average delay performance ($F = 0.5$, $M = 4.5$)

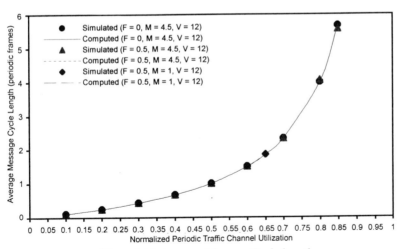

Figure 9.16: Average Message Cycle length

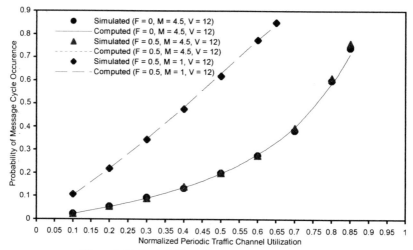

Figure 9.17: Probability of Message Cycle occurrence

9.5 PERFORMANCE COMPARISON

The performance of generalized protocol using RSA is compared with the same protocol using TDMA and conventional polling for a 20-user network. The operational aspects of the three protocols are defined in Table 9.2.

As shown in Figure 9.18, the channel utilizations of all three protocols are highly efficient, even at high network load. This suggests that the performance of the generalized protocol (which uses a random access scheme in the Contention Cycle) does not deteriorate considerably at high load. Figure 9.19 shows that the average delay for the generalized protocol using RSA is better than the other two protocols except when the network is very heavily loaded.

Note that if more users are added, the average delay characteristic for generalized protocol using RSA is expected to outperform the other two protocols, particularly at low network load. This can be explained as follows. For the generalized protocol using TDMA, more time slots will be required to cater for the increased number of users. In the case of conventional polling, more go-ahead messages will be required. Under low network load, since many users will be inactive, these additional time slots or go-ahead messages reduce the channel utilization and increase the average delay.

Table 9.2: Comparison of multiple access protocols

Multiple Access Protocol	Description
Genaralized Protocol using RSA	As described in Section 9.1.
Genaralized Protocol using TDMA	One minislot is specifically assigned to each user for the transmission of reservations.
Conventional Polling	The base station polls each user sequentially using polling signals with zero overheads. A ready user sends a message when polled. An idle user sends a go-ahead message equivalent to the size of a minislot.

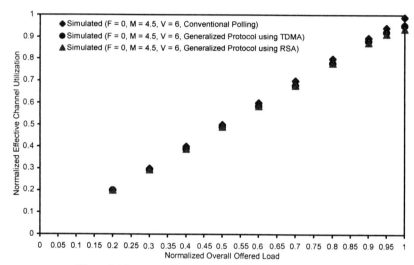

Figure 9.18: Normalized channel utilization comparison

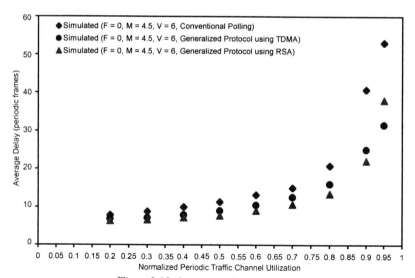

Figure 9.19: Average delay comparison

9.6 TRAFFIC LOAD BALANCING

Due to the high capacity requirement, deployment of broadband wireless networks can be achieved in the form of micro- or pico-cellular networks where communication is partitioned into clusters of small coverage areas (radio cells) with frequency channels that can be reused concurrently in distant locations. Each of these clusters utilizes the entire allocated spectrum (bandwidth) and allows mobile devices to operate at low transmit power. To support multimedia traffic, it is very likely that certain cells will encounter significantly more traffic than others. Furthermore, this traffic imbalance varies with time and is dependent on several factors such as the size of the cells, the user density per cell, and the degree of mobility of the users.

Clearly, the use of small cells with non-uniform traffic implies that fast adaptation to temporal and spatial traffic variations across different cells is essential to network operation. The overall bandwidth utilization can be improved considerably if suitable traffic management or load balancing schemes are implemented [8]. The aim of traffic load balancing is to direct traffic from one cell to another according to the traffic volume in each cell. Users are not restricted to one cell but may join adjacent cells whenever congestion arises. In this way, requests for new connections can be accommodated without severely compromising the quality of service (QoS) requirements of existing connections.

Although dynamic channel allocation, channel borrowing, and channel segregation have the focus of active research for many years, these load balancing methods are complex and may result in excessive delays for interactive applications. Since these methods are not incorporated directly into the access protocol, they may not be responsive to sudden changes in the traffic load.

For the generalized protocol, the feasibility of sharing traffic load relies on the fact that cell boundaries are seldom well-defined due to the complex interaction between traffic intensity and radio propagation characteristics, thus resulting in partial overlaps among cell coverage areas (see Figure 9.20). In fact, cell overlap is often considered in cellular network design. Otherwise, when the subscriber load is increased and cell sizes are reduced, dead spots can be created, thereby causing dropped connections. Thus, it is common to let lightly loaded base stations support adjacent, heavily loaded base stations even though it is obvious that such base stations do not provide the strongest signal most of the time.

As the number of reuse frequencies increases, all users in every cell can theoretically be connected to one or more adjacent base station (see Figure 9.21). This is equivalent to having two radio cells completely overlapping one another (i.e., 100% overlap). However, overlapping areas generally tend to occur at the periphery areas of radio cells. Thus, mobile users that are distant from the base station benefit most from load balancing since they are most likely to be in communication range of nearby base stations. The penalty for achieving traffic load balancing in this manner is increased interference (in particular, co-channel interference) due to connections being carried in cells other than their normal cells. In the case of Figure 9.21, there will also be less available bandwidth for individual cells compared to Figure 9.20.

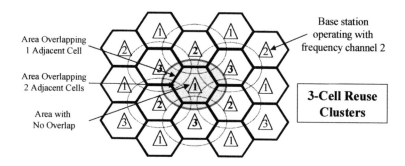

Figure 9.20: Overlapping coverage in cellular systems (3-cell reuse clusters)

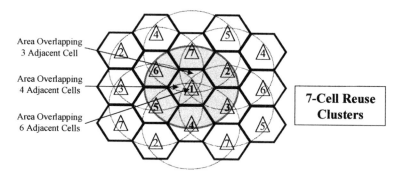

Figure 9.21: Overlapping coverage in cellular systems (7-cell reuse clusters)

For the RSA protocol, it has been shown analytically in Section 9.3.2 that if a minimum of two or more minislots is assigned, then the protocol can always achieve stable non-zero throughput. However, this throughput may not be optimized, especially when there is a big mismatch between the number of minislots and the number of ready users. It will be better if the number of minislots in a Contention Cycle (V) adapt to the number of ready users (or equivalently, the traffic load). Specifically, V should increase when the number of ready users is large. This reduces the transmission probability (p) in a minislot, thereby reducing the number of collisions and increasing the corresponding throughput. On the other hand, when the number of ready users is small, V should decrease so that the average delay is minimized under conditions of low traffic load. However, the decrease in V is subject to a minimum since when $V = 1$, zero throughput results if a collision occurs. Thus, to control V (and hence p) dynamically, an estimation of the number of ready users is mandatory. The main problem here is that the number of ready users is an unknown variable that changes with time.

As explained in Section 9.3.2, numerous collision resolution and dynamic control algorithms have been reported over the years. Although most of the proposed algorithms possess the desirable property of throughput stability, they rely on channel feedback information and require individual users (both ready and idle) to be able to observe outcomes of the broadcast channel and make consistent decisions. To do so, the user must typically be capable of determining after transmission of a packet, whether there have been zero (i.e., idle channel), one (i.e., success), or multiple packets (i.e., collision). The latter requirement is not always viable in an unpredictable wireless channel since a receiver may easily misinterpret the actual outcome of a transmission. Specifically, it can be hard to decipher between an idle and a collision and between a collision and a transmission error. As a result, some of these algorithms are liable to suffer from problems of deadlock arising from erroneous channel feedback when implemented in a wireless environment. Furthermore, such control schemes normally do not cater for overloaded situations where the total arrival rate exceeds the channel capacity. The stability control method applied to the generalized protocol protects the channel from unstable behavior while optimizing channel efficiency and performance during normal operating conditions. In addition, it enables traffic load in a cell to be spread out within a group of cells when that particular cell becomes either heavily utilized or overloaded.

9.6.1 Protocol Description

The beacon signal serves a crucial function here. All beacon signals from individual base stations in the network operate at a common frequency channel (reasons for doing this will be explained shortly). Each beacon signal contains general signaling information (e.g., paging requests, handoff requests). In addition, it specifies the identity and location area of the base station as well as the frequency channel operating within the cell. Users will then operate on the frequency channel specified by the beacon signal. This channel is selected from a group of frequency channels (which are reused in non-interfering cells) and the selection depends on the level of interference detected by the base station from radio link quality measurements. Alternatively, some kind of slow frequency hopping may be employed where orthogonal frequency channels are randomly chosen by the base station periodically. Thus, the need for tedious frequency planning is removed and the system reconfiguration becomes more flexible.

To avoid conflicts when two or more beacon signals transmit simultaneously, coordination among a cluster of base stations within reuse distance is necessary and this can be achieved in several ways. One method is to allow the base stations within a reuse cluster to schedule their transmissions, either through the backbone network or across the wireless channel using a fixed assignment protocol like TDMA. Each base station in a cluster will be assigned a unique time slot to transmit its beacon signal. For a 3-cell reuse cluster, only three time slots are required to form a TDMA frame. These TDMA frames are repeated periodically. Because the beacon signals are typically short in length, the time slots are also of short durations. Therefore, a beacon signal can be issued quickly whenever its transmission time is due. Such a method, however, requires the base stations to be synchronized. To eliminate this requirement, an alternative is to allow the beacon signals from base stations to contend for channel access (e.g., using ALOHA or CSMA protocols) over the wireless medium. Since the number of contending base stations is considerably smaller than the number of users, contention problems will be rare and like the TDMA method, beacon signals can be issued quickly.

The stability control scheme operates solely on the number of successful transmissions (N_s) in each Contention Cycle (equation 9.24). When the traffic load is low, this process effectively reduces the number of minislots and improves the delay at low traffic load. Obviously, V cannot be allowed to decrease to an extremely low value since it produces an unacceptably high collision probability and a definite unstable condition when $V = 1$. As pointed out earlier, imposing a minimum value for V (denoted by V_{min}) also has the beneficial effect of stabilizing channel throughput. Clearly, a large value for V_{min} will result in a more accurate estimate for the channel load at the expense of a slight increase in average delay. From simulation studies, a minimum value of $V_{min} = 6$ appears to be a reasonable choice for a wide range of traffic loads and user densities while a minimum value of $V_{min} = 3$ is suitable for low to medium traffic loads.

$$V_{new} = \begin{cases} \left\lfloor 1 \middle/ \left(\frac{1}{e}\right) \right\rfloor = 3 & N_s = 0 \\ \left\lfloor N_s \middle/ \left(\frac{1}{e}\right) \right\rfloor = \lfloor N_s \times e \rfloor & N_s \geq 1 \end{cases}$$

$$(9.25)$$

Under heavy traffic, equation 9.25 increases the number of minislots in a Contention Cycle, which in turn, reduces the transmission probability. In doing so, the probability of a successful transmission increases and the throughput performance is correspondingly improved. Since an increase in the number of minislots implies an increase in traffic load, users can deduce traffic intensity simply by observing the number of minislots declared by the beacon signal. Since all beacon signals operate at a common frequency channel, the number of slots declared by a beacon signal allows current users operating in a cell as well as potential users from adjacent cells to estimate the traffic loading.

Although the primary purpose here is to maintain the total offered load at the optimal value of unity, a value lower than $1/e$ may be chosen by the network designer if the expected number of users supported per cell is large. The operation of the load balancing scheme is illustrated in Figure 9.22.

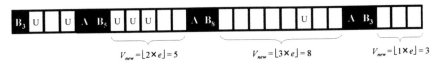

Figure 9.22: Operation of the load balancing scheme

A small value for N_t need not necessarily mean low traffic intensity. On the contrary, it can imply possible channel saturation when V_{new} cannot increase fast enough to meet short-term traffic overloading. It is not possible to distinguish between these two conditions since the proposed control scheme does not make use of collision feedback. Even if a cell is not overloaded but has high channel utilization, a user may still encounter excessive collisions, resulting in long delays and high connection/handoff failure rates.

To reduce the occurrence of cell overloading and minimize collisions, a user with a collided transmission in the Contention Cycle may resort to traffic load balancing. This feature will now be described. A user first transmits a new reservation/bursty frame to the base station of its present cell (i.e., the cell which the user is currently residing). At the same time, the user examines the number of minislots defined by the beacon signals of adjacent cells with adequate signal strength. If the attempted request is unsuccessful and if the number of minislots in one of the neighboring cells is less than the present cell (i.e., the neighboring cell has a lower traffic load than the present cell), the request will be re-attempted in that adjacent cell. Otherwise, the retransmission will have to be attempted in the present cell. The flowchart for this algorithm is shown in Figure 9.23. If two or more adjacent cells are available, the user will either choose the cell with the least number of minislots or randomly select a cell should all available adjacent cells have the same number of slots. If all mechanisms fail after a number of retries, the request will be discarded. By making users attempt retransmissions in another cell, contention in a congested cell is reduced. This improves the probability of success for new connections and the throughput in that cell is correspondingly increased. Thus, more ready users can be accommodated (i.e., a capacity increase).

To summarize, the base station controls the number of minislots according to the amount of network load estimated. When the network loading is light, short Contention Cycles dominate and bursty frames or reservations get transferred across the network quickly. Under heavy traffic or overload conditions, the overall bandwidth is efficiently utilized when ready users exploit the extra capacity offered by adjacent cells with low channel utilizations. Thus, the proposed protocol adapts automatically to a time-varying traffic load spatially.

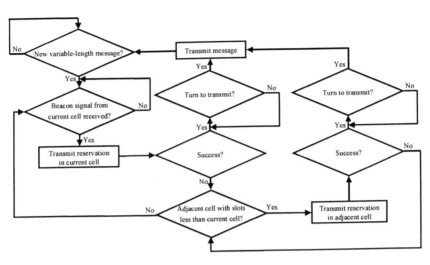

Figure 9.23: Flowchart for user with periodic traffic using traffic load balancing

9.6.2 Simulation Results

In order to reduce the number of parameters required for analysis, the simulation assumed two cells (Current Cell and Adjacent Cell) with 50% and 100% overlapped coverage areas. There are 20 stationary users in either cell and the number of users in the overlapping region is proportional to the amount of overlap. Users in Current Cell generate reservations at a rate ten times greater than users in Adjacent Cell, which carries solely bursty traffic. The length of a minislot is fixed at 1/12th of a periodic frame and the number of retransmissions is not subjected to any limit (i.e., there are no discards). The case without traffic load balancing is simulated using only Current Cell. The general assumptions enumerated in Section 9.3 still apply. As the issue of channel utilization is of prime concern, the effect of selecting a cell with the best signal strength is neglected.

Figures 9.24 to 9.27 depict the average delay and the average number of retransmissions in Current Cell respectively. It can be seen that with traffic load balancing, high channel utilizations can be maintained at the expense of a very small number of retransmissions. This translates to a significant reduction in connection/handoff failure rate and average delay. Included in the diagram is the idealized case where the base station has perfect knowledge of the actual number (but not the identities) of ready users and can therefore allocate the exact number of minislots that optimizes performance. Clearly, the proposed stability control scheme with traffic load balancing performs creditably when compared to the ideal case. The average channel utilization in Adjacent Cell is shown in Figure 9.28. It can be observed that a substantial gain in total admitted traffic can be obtained by handing excess traffic over to adjacent cells with low traffic concentrations. This enables the proposed protocol to exercise congestion control and graceful degradation in the presence of traffic overload.

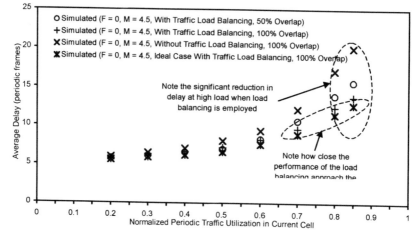

Figure 9.24: Average delay performance in Current Cell ($V_{min} = 3$)

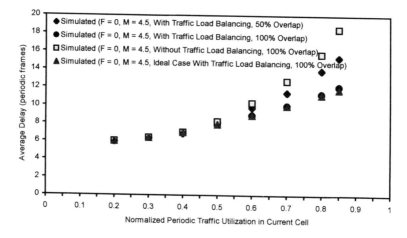

Figure 9.25: Average delay performance in current cell ($V_{min} = 6$)

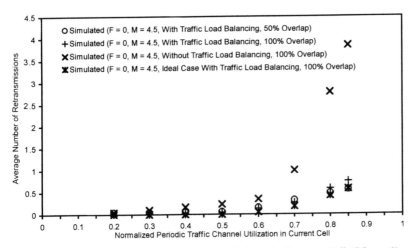

Figure 9.26: Average number of retransmissions in Current Cell ($V_{min} = 3$)

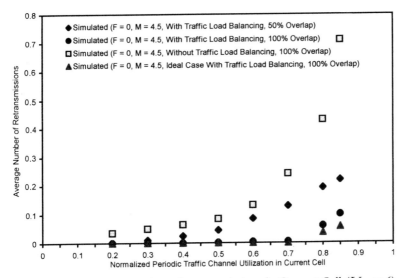

Figure 9.27: Average number of retransmissions in Current Cell ($V_{min} = 6$)

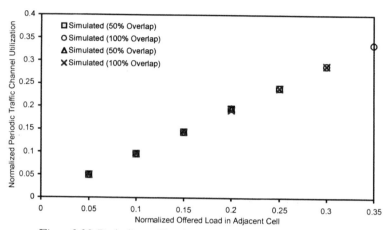

Figure 9.28: Periodic traffic channel utilization in Adjacent Cell

9.7 OPNET SIMULATION MODELS

OPtimized Network Engineering Tools (OPNET) is a stochastic, event-driven network simulator that implements protocol specifications based on finite state diagrams. These state diagrams are modeled graphically with conditions and actions specified on each transition. Verification of the protocol is carried out by a state-space exploration performed automatically by the program. Such an approach enjoys certain desirable properties (e.g., freedom from deadlocks and unspecified exceptions).

OPNET employs a hierarchical approach to the modeling of packet-based networks. At the lowest level, the user creates a process model using a schematic capture tool that embodies the functionality and logic flow of the system/protocol being modeled. The process is specified using a state transition diagram within a graphical user interface. Each state performs specified executives (actions) which are coded as C statements. The resulting state transition diagram defines the logical flow of the protocol in response to discrete-time events that take place. At the second level in the modeling hierarchy, the process model is assigned to a module such as a processor or a queue. The processors or queues are then interconnected to other functional blocks such as transmitters, receivers or generator modules. The entire collection is called a node model and defines the data flow within the node domain of the model. The third level is the network domain where the nodes are interconnected and links are given parameters such as propagation delay, error properties, etc. and the physical topology of the network is generally specified. Finally, the network model is linked to produce an executable file. During this process, probes can be specified to gather data at the nodes and these statistics may be pre-defined or user-defined. Thus, there is complete flexibility to model a wide range of systems and to gather extensive data sets for analysis. More details can be found in [KATZ98].

9.8 OPNET MODELS FOR GENERALIZED PROTOCOL

The network model consists of a single base station communicating with multiple users using a common broadcast medium.

9.8.1 Node Model for Base Station

The node model for the base station consists of a queue module, a transmitter module and a receiver module (Figure 9.29). The transmitter and receiver modules are connected to four subchannels whose functions are listed in Table 9.3. These subchannels operate sequentially, not independently. They are required to handle unwanted stream interrupts when transmissions are inadvertently sent back to the users who broadcast them. In addition, these subchannels allow the implementation of the process models to be simplified considerably.

Table 9.3: Functions of subchannels

Subchannel	Function
0	Transmission of beacon signals from the base station to users
1	Transmission of reservations/bursty frames from users to the base station
2	Transmission of acknowledgment signals from the base station to users
3	Transmission of messages from users with successful reservations to the base station

9.8.2 Node Model for User with Periodic Traffic

The node model for a user with periodic traffic consists of a traffic source module, a processor module, a transmitter module and a receiver module (Figure 9.30). The node model for users with bursty traffic is implemented in a similar manner.

9.8.3 Process Model for Base Station

In the state diagrams, an unforced state represents a resting state in which the simulation process will pause and wait for a discrete-time event (interrupt) before beginning execution again. When an event occurs, an unforced state may change to another state. A forced state is traversed by the simulation process without any passage of time and serves mainly as a convenient branching point for flow-diagram type of protocol specification. A forced state requires an immediate transition to another state. Five forced states and two unforced states are defined for the base station (Figure 9.31).

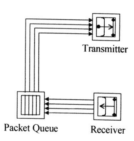

Figure 9.29: Node model for the base station

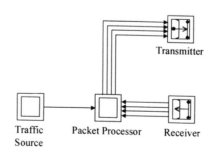

Figure 9.30: Node model for user with periodic traffic

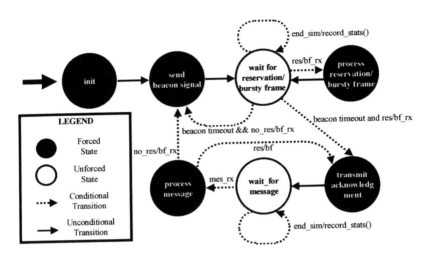

Figure 9.31: Process of the base station

Init Forced State

The global variables that keep track of the statistics of frames received are initialized here. These include variables for successfully received bursty/periodic traffic as well as the total number of frames transmitted.

Send_Beac_Sig Forced State

A broadcast beacon signal is sent to all users. A timeout equivalent to the duration of the total number of random access time slots in a Contention Cycle is scheduled. On expiry of the timeout, base station will proceed to send acknowledgment signals.

Wait_for_Res_Bf Unforced State

This resting state is entered immediately after the beacon signal is sent. The base station waits for reservations or bursty frames to be transmitted by the users.

Process_Res_Bf Forced State

This state is entered when a successful reservation or bursty frame arrives at the base station. The bursty frames and reservations are inserted into a common queue. Since bursty frames will be acknowledged first, they are inserted ahead of reservations. The average delay for each bursty frame received is also computed here.

Send_Poll_Sig Forced State

The identities of the users that have successfully transmitted a reservation or bursty frame are obtained. Bursty frames are acknowledged first. Users with successful reservations are then polled.

Wait_for_Message Unforced State

The base station waits for a message from a user with periodic traffic or an acknowledgment frame from a user with bursty traffic to arrive after sending the acknowledgment or polling signal.

Process_Message Forced State

On arrival of a message, statistics such as its size and delay are determined and written into the global variables. Acknowledgment frames are discarded. The global variables that relate to the number of bursty frames received are also updated.

9.8.4 Process Model for User with Periodic Traffic

The OPNET state diagram for the user with periodic traffic is shown in Figure 9.32. The state diagram for the user with bursty traffic is similar to Figure 9.33. A total of seven states are defined, out of which four are forced states and three are unforced states.

Init Forced State

The discrete uniform distributions for the random transmission of reservations and the modeling of message lengths are loaded.

Idle Unforced State

This is a resting state when no new messages are received.

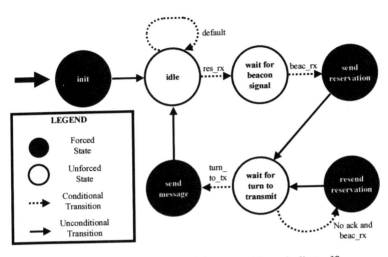

Figure 9.32: Process model of the user with periodic traffic

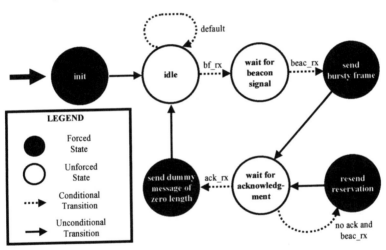

Figure 9.33: Process model of the user with bursty traffic

Wait_for_Beac_Sig Unforced State

This state is entered when a reservation arrives from the traffic source. The user then waits for the arrival of the beacon signal from the base station before transmitting this reservation.

Send_Res Forced State

On receiving a beacon signal from the base station, the user transmits a reservation using the randomized slotted ALOHA protocol. A copy of the reservation is also duplicated in case the transmitted reservation encounters a collision. The global variable that keeps track of the total number of transmitted frames is incremented by one regardless of whether the transmitted reservation is successful or not.

Wait_for_Turn_to_Transmit Unforced State

This state is entered immediately after a reservation has been transmitted.

Resend_Res Forced State

On receiving the beacon signal again, the user knows that the reservation transmitted previously has been unsuccessful and retransmission is now necessary. A copy of the reservation is duplicated in case the retransmitted reservation encounters a collision again. The global variable that keeps track of the total number of transmitted frames is incremented by one.

Send_Mes Forced State

On receiving a polling signal, the user knows that its reservation has been successful and will now transmit its message. The size of this message varies according to the uniform distribution loaded in the *Init* state. The global variable that keeps track of the total number of transmitted frames is incremented by a number equivalent to this size.

9.9 OPNET MODELS TRAFFIC LOAD BALANCING

The network model consists of two base stations communicating with multiple users using a common broadcast medium. The coverage areas of the base stations are denoted as Current Cell and Adjacent Cell respectively. User terminals in Current Cell generate reservations at a rate ten times greater than terminals in Adjacent Cell, which carries solely bursty traffic. In order to reduce the number of parameters needed for analysis, the effect of selecting a cell with the best signal strength is neglected.

9.9.1 Node Model for Base Station

The node model is identical to that for generalized protocol (see Figure 9.29).

9.9.2 Node Model for User with Periodic Traffic

The node model for the node with periodic traffic consists of a queue module, a transmitter module, and a receiver module (Figure 9.34). The bus channel is divided into eight subchannels, four for Current Cell and four for Adjacent Cell.

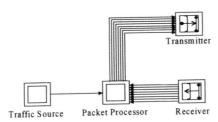

Figure 9.34: Node model for the user with periodic traffic

9.9.3 Process Model for Base Station

The node model is identical to that for generalized protocol (Figure 9.31).

9.9.4 Process Model for User with Periodic Traffic

A total of 16 states are defined out of which five are unforced states while the rest are forced states (Figure 9.35). The functions of these states are summarized as follows. On receiving a new request, a user will first transmit a reservation to the base station of its present cell (i.e., the cell which the user is currently residing). At the same time, the user examines the number of minislots defined by the beacon signal of its adjacent cell. If the attempted request is unsuccessful (i.e., retransmission is necessary) and if the number of minislots in the neighboring cell is less than the present cell (i.e., the neighboring cell has a lower traffic load than the present cell), the request will be re-attempted in that adjacent cell.

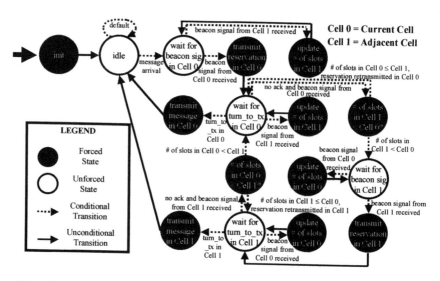

Figure 9.35: Process model of a user with periodic traffic performing traffic load balancing in Current Cell

9.10 SIMULATION TIME

Any simulation will involve a tradeoff between confidence in the results and computational overhead. Initially, the simulation time is varied between 1 ms and 8 seconds to determine the trends in the results, if any, and their dependence on the length of run time. It was concluded that extending the simulation time had no effect on the trends in the data gathered, except that the data sets were longer. Different random seeds were used to improve the reliability of the simulation results.

9.11 FORMAL VERIFICATION OF GENERALIZED PROTOCOL

To verify that a protocol is correct is to determine whether it performs the intended behavior. Numerous verification methods have been proposed and applied to the verification and construction of protocols [HOLT91], [LAM84]. These methods can be formal (as presented here) or automatic. Two popular methodologies for the automatic verification of protocols are Finite State Machines and Petri Nets.

A Finite State Machine (FSM) defines what actions a process is allowed to take, which events it expects to happen and how it will respond to these events. When a protocol is defined, unacceptable sequences of states of the FSM can be detected by a monitor. Reachability analysis can then be used to verify the correctness of the protocol. A state transition diagram summarizes the definition of a FSM. It represents all the states of the FSM together with arrows between states that represent the possible transitions. Unique sets of actions must be specifiable upon both entering and leaving a state. A Petri Net (PN) specifies the evolution of tokens in places connected to transitions by arrows. The configuration of tokens in the places is called the marking of the PN. The set of reachable markings can also be analyzed.

A formal proof on the correctness of the generalized protocol is given. It is assumed that each transmission is received error-free with a probability of at least $q > 0$.

9.11.1 Correctness

Correctness criteria are often classified under safety and liveness. An algorithm is safe if it never produces an incorrect result, which in the case of generalized protocol means never releasing a bursty frame or message out of the correct order to the base station. An algorithm is live if it can continue to produce results indefinitely (i.e., a live algorithm can never enter a deadlock condition from which no further progress is possible). For generalized protocol, liveness refers to the ability to be able to receive transmissions from the users and release them at the base station.

9.11.2 Safety

The safety property for generalized protocol is self-evident from the FCFS assumption in that the $(i + 1)^{th}$ bursty frame/message is accepted by the user only after the ith bursty frame/message has been successfully transmitted.

9.11.3 Liveness

To verify the liveness property, consider first a user that transmits a bursty frame i at time t_1 (Figure 9.36a). Let t_2 be the time at which this frame is received correctly by the base station. Let $t_2 = \infty$ if this event never occurs. Similarly, let t_3 be the time at which the send sequence number (SN) at the user is increased to $i + 1$ and let $t_3 = \infty$ if this never occurs. We will show that $t_1 < t_2 < t_3$ and that t_3 is finite. This is sufficient to demonstrate liveness since, by induction, each bursty frame will then be transmitted with finite delay. Let RN(t) be the value of the variable RN (receive sequence number) at the base station as a function of time and let $SN(t)$ be the corresponding value of SN at the user. We shall adopt the convention that RN indicates the sequence number of the next expected transmission. Since $SN(t)$ and $RN(t)$ are non-decreasing in t and since $SN(t)$ is the largest request number received from the base station up to time t, $SN(t) \le RN(t)$. Using the safety property, $RN(t_1) \le i$. Because $SN(t_1) = i$, it follows that $SN(t_1) = RN(t_1) = i$. By definition of t_2 and t_3, $RN(t)$ is incremented to $i + 1$ at t_2 and $SN(t)$ is incremented to $i + 1$ at t_3. Using the fact that $SN(t) \le RN(t)$, it follows that $t_2 < t_3$.

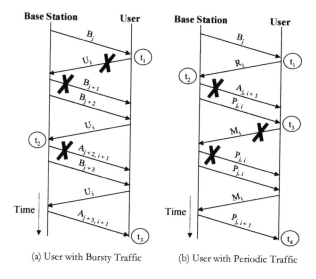

(a) User with Bursty Traffic   (b) User with Periodic Traffic

Figure 9.36: Line diagrams for the verification of the generalized protocol

The user transmits bursty frame i repeatedly, with finite delay between successive transmissions, from t_1 until it is first received error-free at t_2. Since there is a probability $q > 0$ that each retransmission is received correctly and retransmissions occur within finite intervals, an error-free reception eventually occurs and t_2 is finite. A similar argument can be applied to the signals transmitted by the base station to prove that t_3 also occurs eventually. Thus, the interval from t_1 to t_3 is finite and the algorithm is live. Note that if the number of minislots (V) in a Contention Cycle is fixed at one and a collision occurs, then generalized protocol will fail the liveness test.

A similar procedure can be employed to verify the liveness property for a user with periodic traffic (Figure 9.36b). It can be shown that $t_1 < t_2 < t_3 < t_4$ and that t_4 is finite.

SUMMARY

A broadband multiple access strategy described in this chapter achieves flexible bandwidth sharing by allowing mobile users to seize variable amounts of bandwidth on demand. Unlike most random access protocols, it has been shown that this contention-based multiple access technique achieves stable throughput. The new scheme is able to handle an integrated mix of multimedia traffic and operate robustly in an unreliable wireless transport medium. It is also responsive to frequent changes in the traffic load and has a unique capability to distribute traffic load autonomously among different cells in the network. This load balancing feature is a major QoS feature that automatically helps to prevent network overload at the medium access control (MAC) level.

The dynamic characteristics of the RSA protocol that is commonly employed in random access schemes has also been analyzed. A Markov chain for the RSA protocol has been developed which is complemented by a drift analysis under different estimated loads. The estimated backlog is a constant multiple of the actual backlog. Different values of the multiplier correspond to under-estimates or over-estimates. It is found that with the standard single buffer model that the scheme avoids bistability of drift. Results also show that unlike many random access schemes, RSA possesses exactly one equilibrium point even when the channel load estimate is inaccurate. From a practical standpoint, this means that the performance of the RSA scheme degrades gracefully with increased estimation error. However, any under-estimation in the channel load increases the equilibrium backlog (and reduces the average throughput) at a rate faster than when the channel load is over-estimated. In addition, the protocol can achieve stable, non-zero channel throughput by assigning two or more slots during a contention cycle. This eliminates the need for channel feedback information.

An analytical approach based on queuing theory is employed to gain some insight into the behavior of the generalized protocol. This approach appears to be accurate in predicting its performance for both bursty and periodic traffic types. The analytical results are verified by network simulation.

BIBLIOGRAPHY

[BERT92]   Bertsekas, D. and Gallager, R., *Data Networks*, Prentice Hall, 1992.
[BING99a] Bing, B. and Subramanian, R., "Enhanced Reserved Polling Multiaccess Technique for Multimedia Personal Communication Systems", *ACM/Baltzer Journal on Wireless Networks*, Vol. 5, No. 3, May 1999, pp. 221 – 230.
[BING99b] Bing, B. and Subramanian, R., "A Multiaccess Technique for Broadband Wireless Local Networks", *Computer Networks*, Vol. 31, No. 20, November 1999, pp. 2153 – 2169.
[BING99c] Bing, B., "Simulation of Broadband Multiple Access Protocols for Wireless Networks", *Proceedings of the 7th International Symposium on Modeling, Analysis and Simulation of Computer and Telecommunication Systems (MASCOTS '99)*, College Park, Maryland, 24 – 28 October 1999, pp. 92 – 100.
[BING00]   Bing, B., "Stabilization of the Randomized Slotted ALOHA Protocol Without the Use of Channel Feedback Information", *IEEE Communications Letters*, to appear.
[CARL75]   Carleial, A. and Hellman, M., "Bistable Behavior of ALOHA-Type Systems", *IEEE Transactions on Communications*, Vol. COM-23, April 1975, pp. 401 – 410.
[EKLU86]   Eklundh, B., "Channel Utilization and Blocking Probability in a Cellular Mobile Telephone System with Directed Retry", *IEEE Transactions on Communications*, Vol. COM-34, No. 4, April 1986, pp. 329 – 337.

[FAYO77]   Fayolle, G., Gelenbe, E. and Labetoulle, J., "Stability and Optimal Control of the Packet Switching Broadcast Channel", *Journal of the Association for Computing Machinery*, Vol. 24, No. 3, July 1977, pp. 375 – 386.

[FERG75]   Ferguson, M., "On the Control, Stability and Waiting Time in a Slotted ALOHA Random-Access System", *IEEE Transactions on Communications*, Vol. COM-25, November 1975, pp. 1306 – 1311.

[HAJE82a] Hajek, B. and van Loon, "Decentralized Dynamic Control of a Multiaccess Broadcast Channel", *IEEE Transactions on Automatic Control*, Vol. AC-27, No. 3, June 1982, pp. 559 – 569.

[HAJE82b] Hajek, B., "Hitting-Time and Occupation-Time, Bounds Implied by Drift Analysis with Applications", *Advances in Applied Probability*, Vol. 14, pp. 502 – 525, 1982.

[HANS79]   Hansen, L. and Schwartz, M., "An Assigned-Slot Listen-Before-Transmission Protocol for a Multiaccess Data Channel", *IEEE Transactions on Communications*, Vol. COM-27, No. 6, pp. 846 – 857, June 1979.

[HOLT91]   Holzmann, G., *Design and Validation of Computer Protocols*, Prentice Hall, 1991.

[JENQ81]   Jenq, Y., "Optimal Retransmission Control of Slotted ALOHA Systems", *IEEE Transactions on Communications*, Vol. COM-29(6), June 1981, pp. 891 – 894.

[KARL89]   Karlsson, J. and Eklundh, B., "A Cellular Mobile Telephone System with Load Sharing – An Enhancement of Directed Retry", *IEEE Transactions on Communications*, Vol. COM-37, No. 5, May 1989, pp. 530 – 539.

[KELL85]   Kelly, F., "Stochastic Models of Computer Communication Systems", *Journal of the Royal Statistical Society (Series B)*, 47(3), 1985, pp. 379 - 395.

[LAM84a]   Lam, S., *Principles of Communication and Networking Protocols*, IEEE Computer Society Press, 1984.

[LAM84b]   Lam, S., "Formal Methods for Protocol Verification and Construction", appearing in [LAM84a], pp. 463 – 466.

[METC73]   Metcalfe, R., "Steady State Analysis of a Slotted and Controlled ALOHA System with Blocking", *Proceedings of the 6th Hawaii International Conference on Systems Sciences*, January 1973, pp. 375 – 378.

[RIVE87]   Rivest, R., "Network Control by Bayesian Broadcast", *IEEE Transactions on Information Theory*, Vol. IT-33, No. 3, May 1987, pp. 323 - 328.

[ROM90]   Rom, R. and Sidi, M., *Multiple Access Protocols: Performance and Analysis*, Springer-Verlag, 1990.

[WATA95] Watanabe, F., Buot, T., Iwama, T. and Mizuno, M., "Load Sharing Sector Cells in Cellular Systems", *Proceedings of the IEEE PIMRC*, 1995, pp. 547 – 551.

[WOOD94] Woodward, M., *Communication and Computer Networks*, IEEE Computer Society Press, 1994.

Appendix

A QUEUING THEORY PRIMER

Several basic queuing theory concepts needed for understanding the analysis carried out in several parts of the book are discussed in this appendix. A list of symbols is defined in Table A.1.

A.1 THE POISSON PROCESS

Under a Poisson packet arrival process, the probability of n arrivals for a given time interval t s and an arrival rate of λ packets/s is given by:

$$P_n = \frac{(\lambda t)^n e^{-\lambda t}}{n!} = \frac{G^n e^{-G}}{n!}$$

(A.1)

where $G = \lambda t$.

Table A.1: List of definitions

Symbol	Definition
D	Average delay in system (queue and server)
G	Offered load
N_q	Number of packets in queue
P_n	Probability of n packet arrivals
T	Packet transmission time
V	Server vacation period
W	Waiting time in queue
λ	Packet arrival rate
μ	Packet transmission rate $(=1/T)$
ρ	Utilization of server (i.e., fraction of time server is busy doing work)

248

From equation A.1, the probability of zero, one, and two or more packet arrivals can be obtained easily as follows:

$$P_0 = e^{-G}$$

(A.2)

$$P_1 = Ge^{-G}$$

(A.3)

$$P_{\geq 2} = 1 - e^{-G} - Ge^{-G}$$

(A.4)

If $G = 1$, then $P_0 = P_1 = 1/e \approx 0.368$ and $P_{\geq 2} = 1 - 2/e \approx 0.264$. In terms of slotted ALOHA (Section 5.2.2), this means on the average, 36.8% of the slots are idle, another 36.8% contain successful packet transmissions while the rest of the slots contain collisions.

A.2 LITTLE'S THEOREM

In simple terms, Little's theorem states that:

Average number of packets in system = average delay × throughput

(A.5)

This theorem applies to any system with a well-defined boundary. For example, in a single server queue (Figure A.1), the system can comprise the queue or the queue plus server.

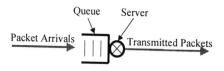

Figure A.1: A single server queue

A.3 SINGLE SERVER QUEUE

The average packet transmission time in a single server queue is given by:

$$\overline{T} = \frac{1}{n}\sum_{i=1}^{n} T_i$$

(A.6)

where T_1, T_2 ... T_n are samples of the random variable T.

Variation of the packet transmission time is modeled using its second moment and is given by:

$$\overline{T^2} = \frac{1}{n}\sum_{i=1}^{n} T_i^2$$

(A.7)

A new packet may arrive at a time when another packet is already transmitting. The average residual time represents the average remaining time for the current packet to complete transmission and is given by:

$$\overline{R} = \frac{\overline{T^2}}{2\overline{T}}$$

(A.8)

If the system is work conserving (i.e., system does not turn away or lose packets), then the server utilization is given by:

$$\rho = \frac{\lambda}{\mu} = \lambda\overline{T}$$

(A.9)

A.3.1 M/G/1 Queue

For a work conserving single server queue with Poisson arrivals and arbitrary packet transmission time,

$$W = N_q \overline{T} + \rho \overline{R} = (\lambda W)\overline{T} + (\lambda \overline{T})\frac{\overline{T^2}}{2\overline{T}} = \rho W + \frac{\lambda \overline{T^2}}{2}$$

(A.10)

$$\therefore W = \frac{\lambda \overline{T^2}}{2(1-\rho)}$$

(A.11)

$$D = \frac{\lambda \overline{T^2}}{2(1-\rho)} + \overline{T}$$

(A.12)

For a single server queue with vacations [BERT92], the server goes on vacation for a random period (V) when all packets in a busy period have been serviced. Thus, the vacation period occurs with a probability $(1 - \rho)$, the probability when there are no packets in the system. The waiting time then becomes:

$$W = N_q \overline{T} + \rho \overline{R} + (1-\rho)\frac{\overline{V^2}}{2\overline{V}} = \rho W + \frac{\lambda \overline{T^2}}{2} + (1-\rho)\frac{\overline{V^2}}{2\overline{V}}$$

(A.13)

$$\therefore W = \frac{\lambda \overline{T^2}}{2(1-\rho)} + \frac{\overline{V^2}}{2\overline{V}}$$

$$(A.14)$$

A.3.2 M/M/1 Queue

If the packet transmission time is exponentially distributed,

$$\overline{T^2} = 2(\overline{T})^2$$

$$(A.15)$$

From equation A.11,

$$W_{M/M/1} = \frac{\rho \overline{T}}{1-\rho}$$

$$(A.16)$$

From equation A.12,

$$T_{M/M/1} = \frac{\rho \overline{T}}{1-\rho} + \overline{T} = \frac{\overline{T}}{1-\rho} = \frac{\rho}{1-\rho}\frac{1}{\lambda}$$

$$(A.17)$$

A.3.3 M/D/1 Queue

If the packet transmission time is fixed,

$$\overline{T^2} = \overline{T}^2$$

$$(A.18)$$

252

$$W_{M/D/1} = \frac{\rho\overline{T}}{2(1-\rho)} = \frac{1}{2}W_{M/M/1}$$

(A.19)

$$T_{M/D/1} = \frac{\rho\overline{T}}{2(1-\rho)} + \overline{T} = \frac{\overline{T}(2-\rho)}{2(1-\rho)} = \frac{\rho(2-\rho)}{2(1-\rho)}\frac{1}{\lambda} = \left(1-\frac{\rho}{2}\right)T_{M/M/1}$$

(A.20)

Thus, packets with fixed lengths, being less random, experience less delay than packets with exponential lengths.

A.4 CONSERVATION LAWS

Work conservation in a system gives rise to a conservation law for average waiting times (i.e., a linear relation between average waiting times). The conservation law does not depend on the order of packet transmission as long as packets are not pre-empted.

The polling multiple access technique (Section 4.4), is not work conserving since the server remains idle (during the times when polling messages are sent) even though packets may be present in the system. However, if no overheads are assumed, then some packet is always being served if there are packets in the system. It is then reasonable to regard the polling system as a distributed M/G/1 queue.

BIBLIOGRAPHY

[BERT92] Bertsekas, D. and Gallager, R., *Data Networks*, Prentice Hall, 1992.
[DAIG92] Daigle, J., *Queueing Theory for Telecommunications*, Addison Wesley, 1992.
[KLEI75] Kleinrock, L., *Queueing Systems Volume 1: Theory*, John Wiley, 1975.

ACRONYMS

2G	Second Generation
3G	Third-Generation
AAL	ATM Adaptation Layer
ABR	Available Bit Rate
ACK	Acknowledgment
ACL	Asynchronous Connectionless
ACTS	Advanced Communications Technologies and Services
ADSL	Asymmetric Digital Subscriber Line
ARQ	Automatic Repeat Request
ARRA	Announced Retransmission Random Access
ATM	Asynchronous Transfer Mode
AWGN	Additive White Gaussian Noise
BER	Bit Error Rate
BMAP	Bit-Map Access Protocol
BPSK	Binary Phase Shift Keying
BRAM	Broadcast Recognition Access Method
BRAN	Broadband Radio Access Network
BTMA	Busy Tone Multiple Access

CAC   Channel Access Control
CBR   Constant Bit Rate
CCI   Co-channel Interference
CCK   Complementary Code Keying
CDMA   Code Division Multiple Access
CEPT   European Conference of Postal and Telecommunications
CLR   Cell Loss Ratio
CSMA   Carrier Sense Multiple Access
CSMA/CA   CSMA with Collision Avoidance
CSMA/CD   CSMA with Collision Detection
CTS   Clear to Send
DAB   Digital Audio Broadcasting
DAMA   Demand Access Multiple Access
DAVIC   Digital Audio-Visual Council
DBPSK   Differential Binary Phase Shift Keying
DCF   Distributed Coordination Function
DCT   Discrete Cosine Transform
DECT   Digital Enhanced Cordless Telecommunications
DFE   Decision Feedback Equalizer
DIFS   DCF Interframe Space
DMT   Discrete Multitone
DQPSK   Differential Quarternary Phase Shift Keying
DQRUMA Distributed Queuing Request Update Multiple Access
DRMA   Dynamic Reservation Multiple Access
DSP   Digital Signal Processing
DSSS   Direct Sequence Spread Spectrum
DS-CDMA Direct Sequence   CDMA
ETSI   European Telecommunications and Standard Institute
EY-NPMA   Elimination Yield Non Pre-emptive Multiple Access
FCC   Federal Communications Commission
FCFS   First-Come-First-Served
FDD   Frequency Division Duplex
FDMA   Frequency Division Multiple Access
FEC   Forward Error Correction
FFT   Fast Fourier Transform
FHSS   Frequency Hopping Spread Spectrum
FSK   Frequency Shift Keying
GFSK   Gaussian Frequency Shift Keying

GMSK	Gaussian Minimum Shift Keying
GPS	Global Positioning System
HDLC	High-level Data Link Control
HiperLAN	High Performance Radio LAN
IEC	International Electrotechnical Commission
IEEE	Institute of Electrical and Electronic Engineers
IETF	Internet Engineering Task Force
IFS	Interframe Space
IMT	International Mobile Telecommunications
IP	Internet Protocol
ISI	Intersymbol Interference
ISM	Industrial, Scientific and Medical
ISO	International Standards Organization
ITU	International Telecommunications Union
ITU-R	ITU – Radio Sector
ITU-T	ITU – Telecommunications Sector
LAN	Local Area Network
LLC	Logical Link Control
LMDS	Local Multipoint Distribution Service
LTE	Linear Transversal Equalizer
MAC	Medium Access Control
MACS	Mixed ALOHA Carrier Sense
MAHO	Mobile Assisted Handoff
MAI	Multiaccess Interference
MF-TDMA	Multi-Frequency TDMA
MKK	Musen-setsubi Kensa-kentei Kyokai (Radio Equipment Inspection and Certification Institute)
MLSE	Maximum Likelihood Sequence Estimator
MLMA	Multilevel Multiple Access
MMAC-PC	Multimedia Mobile Access Communication Systems Promotion Council
MMDS	Multichannel Multipoint Distribution Service
MPEG	Motion Pictures Expert Group
MSAP	Mini-Slotted Alternating Priorities
MTSO	Mobile Telephone Switching Office
NAV	Network Allocation Vector
NCHO	Network Controlled Handoff
NRL	Normalized Residual Lifetime

OFDM	Orthogonal Frequency Division Multiplexing
OQPSK	Offset Quadrature Phase Shift Keying
PCF	Point Coordination Function
PDC	Personal Digital Cellular
PDU	Protocol Data Unit
PIFS	PCF Interframe Space
PN	Pseudonoise
PODA	Priority Oriented Demand Assignment
PRMA	Packet Reservation Multiple Access
QAM	Quadrature Amplitude Modulation
QoS	Quality of Service
QPSK	Quarternary Phase Shift Keying
R-ALOHA	Reservation ALOHA
RAMA	Resource Auction Multiple Access
Ready User	User with data to transmit
RF	Radio Frequency
RSA	Randomized Slotted ALOHA
RTS	Request to Send
RTT	Radio Transmission Technology
SFD	Start Frame Delimiter
SIFS	Short IFS
SAMA	Simple Asynchronous Multiple Access
SCO	Synchronous Connection Oriented
SRMA	Split-Channel Reservation Multiple Access
SRUC	Split Reservation Upon Collision
SSMA	Spread Spectrum Multiple Access
SWAP	Shared Wireless Access Protocol
TDD	Time Division Duplex
TDM	Time Division Multiplex
TDMA	Time Division Multiple Access
TIA	Telecommunications Industry Association
UBR	Unspecified Bit Rate
UMTS	Universal Mobile Telecommunications System
U-NII	Unlicensed National Information Infrastructure
UWC-136	Universal Wireless Communication 136
VBR	Variable Bit Rate
WARC	World Administrative Radio Conference
W-CDMA	Wideband CDMA

| WDM | Wavelength Division Multiplexing |
| WLL | Wireless Local Loop |

ABOUT THE AUTHOR

Benny Bing is currently with the NASA Center for Satellite and Hybrid Communication Networks at the University of Maryland in College Park, Maryland, USA. Previously, he has taught courses related to data communications and computer networking when he was a faculty member with the Department of Electronic and Computer Engineering at Ngee Ann Polytechnic in Singapore. He participated in various satellite communications and networked video conferencing projects while working with Singapore Telecom and AT&T Global Information Solutions.

Benny is active in the area of wireless communications research, contributing over 20 refereed papers in major journals and conferences. His paper on wireless ATM was judged to be one of the best papers at the *1998 IEEE International Conference on ATM*. He is guest editor on "Multiple Access for Broadband Wireless Networks" for the *IEEE Communications Magazine*. He is also the author of the best-selling book *High-Speed Wireless ATM and LANs* published by Artech House, Boston, February 2000. He continues to serve as a member of the international advisory committee for the *IEEE International Conference on ATM*, the *IEEE International Conference on Network*, as well as a technical session chair for several international conferences. His lecture on wireless LANs was featured in the *2000 Online Symposium for Electronic Engineers*.

Benny obtained his undergraduate and graduate degrees from the Nanyang Technological University in Singapore, all in Electrical and Electronic Engineering. He is a recipient of the Lockheed Martin Global Telecommunications Graduate Fellowship in Communication Networking and the Ngee Ann Polytechnic Undergraduate Scholarship. His main research interests include high-speed wireless networks, optical fiber communications, and queuing theory. Besides his wide interest in communication networks, he is also an avid fan of human computer interface and graphic design. He can be reached at bennybing@ieee.org.

INDEX

A

ALOHA protocols, 77
Adaptive Protocols, 108
Admission Control, 39
Antenna Diversity, 11
Application Adaptation, 39
Attenuation, 7
Automatic Repeat Request, 13

B

Bluetooth, 180
Bursty Traffic, 33
Broadband Services, 35

C

Carrier Sense Multiple Access, 103
CDMA2000, 193
Channel Assignment, 24
Cochannel Interference, 16
Coherence Bandwidth, 4
Collisions, 53
Constant-Envelop Modulation, 17
Correctness, 242

D

Delay Spread, 3
Distributed Coordination Function, 166
DSSS, 121
Doppler Spread, 5
Duplexing, 20

E

Equalization, 11
Error Control, 13

F

Feedback Algorithms, 96
FHSS, 130
Forward Error Correction, 13
Frequency Division Duplex, 20
Frequency Division Multiple Access, 61

G

Gaussian Channel, 9

H

Handoff, 22
Hidden User, 106
High Frequency Propagation, 8
HiperLAN, 172
HomeRF, 179
Hybrid Protocols, 9

I

IEEE 802.11, 165
IMT-2000, 190
Interference Cancellation, 136
Interframe Space, 166
Intersymbol Interference, 2